GEOLOGY AND MINERALOGY RESEARCH DEVELOPMENTS

CONFLICT MINERALS

RESPONSIBLE SOURCING ISSUES AND FACTORS IMPACTING SEC RULE

GEOLOGY AND MINERALOGY RESEARCH DEVELOPMENTS

Additional books in this series can be found on Nova's website under the Series tab.

Additional e-books in this series can be found on Nova's website under the e-book tab.

GEOLOGY AND MINERALOGY RESEARCH DEVELOPMENTS

CONFLICT MINERALS

RESPONSIBLE SOURCING ISSUES AND FACTORS IMPACTING SEC RULE

GARY L. SIMMONS
EDITOR

nova publishers
New York

For permission to use material from this book please contact us:
Telephone 631-231-7269; Fax 631-231-8175
Web Site: http://www.novapublishers.com

Library of Congress Cataloging-in-Publication Data

ISBN: 978-1-63463-326-0

Published by Nova Science Publishers, Inc. † New York

CONTENTS

PREFACE

Since the Dodd-Frank Wall Street Reform and Consumer Protection Act (the Act) was passed in 2010, relevant U.S. agencies have undertaken various activities related to responsible sourcing of conflict minerals from the Democratic Republic of the Congo (DRC) and adjoining countries. In response to the Act, the Department of State (State) and the U.S. Agency for International Development (USAID) developed a strategy in 2011 to address the linkages among human rights abuses, armed groups, and the mining of conflict minerals and are implementing various strategy objectives. This book examines, among other things, the extent to which relevant U.S. agencies have undertaken activities related to responsible sourcing of conflict minerals; and what is known about the status of, and information provided by, stakeholder initiatives focused on responsible sourcing of conflict minerals from the DRC and adjoining countries.

Chapter 1 – Armed groups in eastern DRC continue to commit severe human rights abuses and profit from the exploitation of minerals, according to reports from the United Nations. Congress included a provision in the 2010 Dodd-Frank Wall Street Reform and Consumer Protection Act to address the trade in "conflict minerals"—tin, tantalum, tungsten, and gold. Section 1502 of the Act directed several U.S. agencies to report or focus on issues related to conflict minerals. This report examines, among other things, (1) the extent to which relevant U.S. agencies have undertaken activities related to responsible sourcing of conflict minerals and (2) what is known about the status of, and information provided by, stakeholder initiatives focused on responsible sourcing of conflict minerals from the DRC and adjoining countries. GAO reviewed and analyzed documents and data covering 2003 through 2014. We interviewed representatives from State, USAID, SEC, Commerce, nongovernmental organizations, industry, and international organizations who are cognizant of conflict minerals issues.

Chapter 2 – The eastern part of the DRC has experienced recurring conflicts involving armed groups that have resulted in severe human rights abuses. In addition, armed groups have profited from the exploitation of minerals. In 2010, Congress enacted Section 1502(b) of the Dodd-Frank Wall Street Reform and Consumer Protection Act to address the exploitation of conflict minerals, which include tin, tantalum, tungsten, and gold, and the extreme levels of violence in the DRC. As required by Section 1502(b), the SEC issued a rule in August 2012 that requires companies to disclose their use of conflict minerals and the origin of those minerals. The act requires GAO to report on the rule's effectiveness, among other issues, beginning in 2012 and annually thereafter. Initial company disclosure reports to SEC that would enable GAO to assess the effectiveness of the rule will not be due until May 2014. This report describes, among other issues, (1) factors that may impact whether SEC's rule denies armed groups in the DRC benefits from conflict minerals and (2) information about companies that use conflict minerals and are not required to report to SEC under the rule. GAO reviewed and analyzed documents and interviewed representatives from SEC, the Department of State, the U.S. Agency for International Development, industry associations, NGOs, consulting firms, and international organizations. GAO also analyzed smelter and refiner information. This report does not contain recommendations.

In: Conflict Minerals
Editor: Gary L. Simmons
ISBN: 978-1-63463-326-0

Chapter 1

CONFLICT MINERALS: STAKEHOLDER OPTIONS FOR RESPONSIBLE SOURCING ARE EXPANDING, BUT MORE INFORMATION ON SMELTERS IS NEEDED*

United States Government Accountability Office

WHY GAO DID THIS STUDY

Armed groups in eastern DRC continue to commit severe human rights abuses and profit from the exploitation of minerals, according to reports from the United Nations. Congress included a provision in the 2010 Dodd-Frank Wall Street Reform and Consumer Protection Act to address the trade in "conflict minerals"—tin, tantalum, tungsten, and gold. Section 1502 of the Act directed several U.S. agencies to report or focus on issues related to conflict minerals.

This report examines, among other things, (1) the extent to which relevant U.S. agencies have undertaken activities related to responsible sourcing of conflict minerals and (2) what is known about the status of, and information provided by, stakeholder initiatives focused on responsible sourcing of conflict minerals from the DRC and adjoining countries. GAO reviewed and analyzed documents and data covering 2003 through 2014. We interviewed

* This is an edited, reformatted and augmented version of the United States Government Accountability Office publication, GAO-14-575, dated June 2014.

representatives from State, USAID, SEC, Commerce, nongovernmental organizations, industry, and international organizations who are cognizant of conflict minerals issues.

WHAT GAO RECOMMENDS

GAO recommends that the Secretary of Commerce provide Congress a plan that outlines the steps, with associated timeframes, to develop and report the required information about smelters and refiners of conflict minerals worldwide. Commerce concurred with GAO's recommendation and noted that it will submit a listing of all known conflict minerals processing facilities worldwide to Congress by September 1, 2014.

WHAT GAO FOUND

Since the Dodd-Frank Wall Street Reform and Consumer Protection Act (the Act) was passed in 2010, relevant U.S. agencies have undertaken various activities related to responsible sourcing of conflict minerals from the Democratic Republic of the Congo (DRC) and adjoining countries. In response to the Act, the Department of State (State) and the U.S. Agency for International Development (USAID) developed a strategy in 2011 to address the linkages among human rights abuses, armed groups, and the mining of conflict minerals and are implementing various strategy objectives. The Securities and Exchange Commission (SEC) issued a rule in 2012 requiring certain companies to disclose the source and chain of custody of necessary conflict minerals in their products. However, the Department of Commerce (Commerce) has not yet compiled a list of all conflict minerals processing facilities—smelters and refiners—known worldwide, required by January 2013 by the Act. Commerce cited difficulties with, for example, tracking conflict minerals operations but told GAO that it had completed outreach efforts with the majority of stakeholders. Commerce did not have a plan of action, with associated time frames, for developing and reporting on the list of conflict minerals processing facilities worldwide. Standard practices in program and project management include, among other things, developing a plan to execute specific projects needed to obtain defined results within a specific time frame. An action plan with timeframes could better position

Commerce to report on the status of its efforts to produce a final list to Congress and to hold its personnel accountable for completing activities.

Over the past several years, a number of stakeholders—foreign governments, multilateral organizations, and industry associations, among others—have expanded, or made plans to expand, initiatives focused on responsible sourcing of conflict minerals in the DRC and adjoining countries. These stakeholder initiatives, such as in-region tracing of conflict minerals and development of guidance documents and audit protocols, have grown to include new mine sites, countries, and smelters. For example, the Conflict-Free Smelter Program, an industry-led effort, has expanded from 26 smelters certified as conflict-free in 2013 to 85 smelters as of April 25, 2014 (see table). New stakeholder initiatives are also underway or planned in the region, including the first responsible sourcing initiative in the Congo-Brazzaville. Some initiatives have yielded publicly available information, including data on production of conflict-free minerals and export data. For example, one stakeholder has reported production data for tin, tungsten, and tantalum from three provinces in the DRC and in Rwanda.

Number of Smelters and Refiners in the Conflict-Free Smelter Program as of April 25, 2014

Mineral	Number of smelters/ refiners certified as compliant	Number of smelters/ refiners working toward certification	Total
Tantalum	28	1	29
Tin	13	14	27
Tungsten	1	8	9
Gold	43	2	45
Total	**85**	**25**	**110**

Source: Conflict-Free Sourcing Initiative data, GAO (analysis).

ABBREVIATIONS

BGR	German Federal Institute for Geosciences and Natural Resources
CFSI	Conflict-Free Sourcing Initiative
CFTI	Conflict-Free Tin Initiative
Commerce	United States Department of Commerce

DHS	Demographic and Health Survey
Dodd-Frank Act	Dodd-Frank Wall Street Reform and Consumer Protection Act
DRC	Democratic Republic of the Congo
EICC	Electronic Industry Citizenship Coalition
GeSI	Global e-Sustainability Initiative
EU	European Union
ICGLR	International Conference on the Great Lakes Region
iTSCi	ITRI Tin Supply Chain Initiative
LBMA	London Bullion Market Association
M-23	March 23 Movement
MONUSCO	UN Organization Stabilization Mission in the Democratic Republic of the Congo
NGO	nongovernmental organization
OECD	Organisation for Economic Co-operation and Development
PPA	Public-Private Alliance for Responsible Minerals Trade
SEC	Securities and Exchange Commission
State	United States Department of State
UN	United Nations
USAID	United States Agency for International Development

* * *

June 26, 2014

Congressional Committees

Over the past decade, the United States and the international community have sought to improve security in the Democratic Republic of the Congo (DRC), the site of one of the world's worst humanitarian crises. The International Rescue Committee has previously estimated that since 1998, more than 5.4 million people have died in the DRC as a result of this crisis, which has also destabilized the minerals-rich eastern part of the country, created insecurity, displaced thousands of people, and perpetuated a cycle of poverty. As we previously reported, illegal armed groups and some units of the Congolese national military have committed severe human rights abuses and mass killings and profited from the illegal exploitation of minerals originating in eastern DRC, particularly in the provinces of North Kivu and South Kivu.[1]

Despite some success in efforts to improve security in eastern DRC—notably, the November 2013 defeat of M-23, an illegal armed group,[2] as well as campaigns against other illegal armed groups—the UN reported that as of December 2013, illegal armed groups in eastern DRC continued to pose a threat to security, to be responsible for human rights abuses and displacements of people, and to derive funding from the illicit production, trade, and smuggling of conflict minerals.[3]

Citing the continuing urgency of the humanitarian situation and the need to take action, in July 2010, Congress included in the Dodd-Frank Wall Street Reform and Consumer Protection Act (hereafter referred to as the Act or Dodd-Frank Act) provisions pertaining to trade involving conflict minerals—tantalum, tin, tungsten, and gold.[4] Section 1502 of the Act directed several U.S. agencies—the Securities and Exchange Commission (SEC), the Department of State (State), the U.S. Agency for International Development (USAID), and the Department of Commerce (Commerce)—to take certain actions to implement the Act's conflict minerals provisions, including provisions related to responsible sourcing of conflict minerals from the DRC and the adjoining countries.[5] The Act specifically directed the SEC to promulgate a rule, which SEC issued in August 2012, that requires certain issuers to disclose annually to SEC whether any of such minerals are necessary to the functionality or production of products manufactured, or contracted to be manufactured, by the issuer originated in the DRC or adjoining country and if so, to provide an additional report.[6] The Act also mandated GAO to report, beginning in 2012 and annually thereafter, on the effectiveness of the SEC rule in promoting peace and security in the DRC and adjoining countries, and annually, beginning in 2011, on the rate of sexual violence in war-torn areas of DRC and adjoining countries, among other things.

To address our mandates, since July 2011 we have issued several reports on topics ranging from the rate of sexual violence in the DRC and adjoining countries to SEC's rule and initiatives focused on responsible sourcing of minerals.[7] Because the first SEC-required disclosures of manufacturers' use of conflict minerals were not due to SEC until June 2014, and because sufficient time must elapse to allow the rule's full impact to materialize, we have not yet addressed the effectiveness of SEC's conflict minerals rule as required under the legislation. Moreover, our ability to report on the rule's effectiveness will depend on the availability of information on stakeholder initiatives to, for example, monitor, ensure, or report on responsible sourcing of conflict minerals from the DRC and adjoining countries.[8] Consequently, in this report, we examine

1. the extent to which relevant U.S. agencies have taken actions related to the responsible sourcing of conflict minerals;
2. what is known about the status of, and any information provided by, stakeholder initiatives focused on responsible sourcing of conflict minerals from the DRC and adjoining countries; and
3. any new information about the rate of sexual violence in eastern DRC and three adjoining countries—Burundi, Rwanda, and Uganda—that has become available since we issued our 2013 report.[9]

To address our first and second objectives, we reviewed and analyzed available data covering 2003 through 2014 on conflict minerals from the DRC and adjoining countries and from responsible sourcing initiatives, as well as reports and other documents from relevant U.S. agencies, foreign governments, multilateral organizations, nongovernmental organizations (NGOs), and industry associations. Because the data were not used to support findings, conclusions, or recommendations, we did not assess their reliability. We also interviewed State, USAID, Commerce, and SEC officials as well as stakeholder representatives from international organizations, NGOs, and industry associations who are cognizant of conflict minerals issues. We selected these stakeholders based on their expertise on responsible sourcing issues, because they represented a range of perspectives on conflict minerals, and because we had established contacts with these entities on our last review.[10] In addition, some of the stakeholders we talked to have been working on the ground in the DRC. The stakeholders we spoke with constitute a nongeneralizable sample, and the information we gathered from them cannot be used to infer views of other stakeholders cognizant of conflict minerals issues. To address our third objective, we reviewed and analyzed relevant documents and interviewed representatives from State, USAID, various UN agencies, relevant NGOs, and researchers to identify any new data on sexual violence. Specifically, we followed up with officials from those groups we interviewed for our prior review on sexual violence rates in eastern DRC and adjoining countries.[11] We also conducted Internet literature searches to identify new academic articles containing any additional information available since our 2013 report. See appendix I for a complete description of our scope and methodology.

We conducted this performance audit from September 2013 to June 2014 in accordance with generally accepted government auditing standards. Those standards require that we plan and perform the audit to obtain sufficient, appropriate evidence to provide a reasonable basis for our findings and conclusions based on our audit objectives. We believe that the evidence obtained provides a reasonable basis for our findings and conclusions based on our audit objectives.

BACKGROUND

Brief History of Conflict in the DRC and the Region

The DRC is a vast, mineral-rich nation with an estimated population of about 75 million people and an area that is roughly one-quarter the size of the United States, according to the UN. The map in figure 1 shows the DRC's provinces and adjoining countries.

Since its independence in 1960, the DRC has undergone political upheaval, including a civil war, according to State. In particular, the eastern DRC has continued to be plagued by violence often perpetrated by illegal armed groups and some members of the Congolese national military against civilians. In November 2012, M-23, an illegal armed group, occupied the city of Goma and other cities in eastern DRC and clashed with the Congolese national army. During this time, the UN reported numerous cases of sexual violence against civilians, including women and children, that were perpetrated by armed groups and some members of the Congolese national military.[12] Although M-23 eventually withdrew from the cities, the group's presence in the region continued. In December 2012, the Ugandan president began to broker peace talks, known as the Kampala Dialogue, between M-23 and the DRC government, aimed at reaching a final and principled agreement that ensured the disarmament and demobilization of M-23 and accountability for human rights abuses. The M-23 was defeated in November 2013 by the Congolese national military with support from UN forces. In December 2013, the former M-23 and the DRC each signed individual declarations that, among other things, set out the conditions for the disarmament, demobilization, and reintegration of M23 into Congolese society and called for those responsible for war crimes and crimes against humanity to be held accountable.

Prior to the defeat of M-23, in February 2013, the 11 countries in the region adopted the "Peace, Security and Cooperation Framework for the Democratic Republic of the Congo and the Region."[13] Some of the adjoining countries in the region have also experienced recent turmoil, which has led to flows of large numbers of refugees into the DRC in addition to internally displaced persons.[14] The United Nations High Commissioner for Refugees

(UNHCR) estimated, as of mid 2013, that there were close to 50,000 refugees from the Central African Republic, in addition to over 120,000 refugees from other countries, as well as around 2.6 million internally displaced persons living in camps or with host families in the DRC.

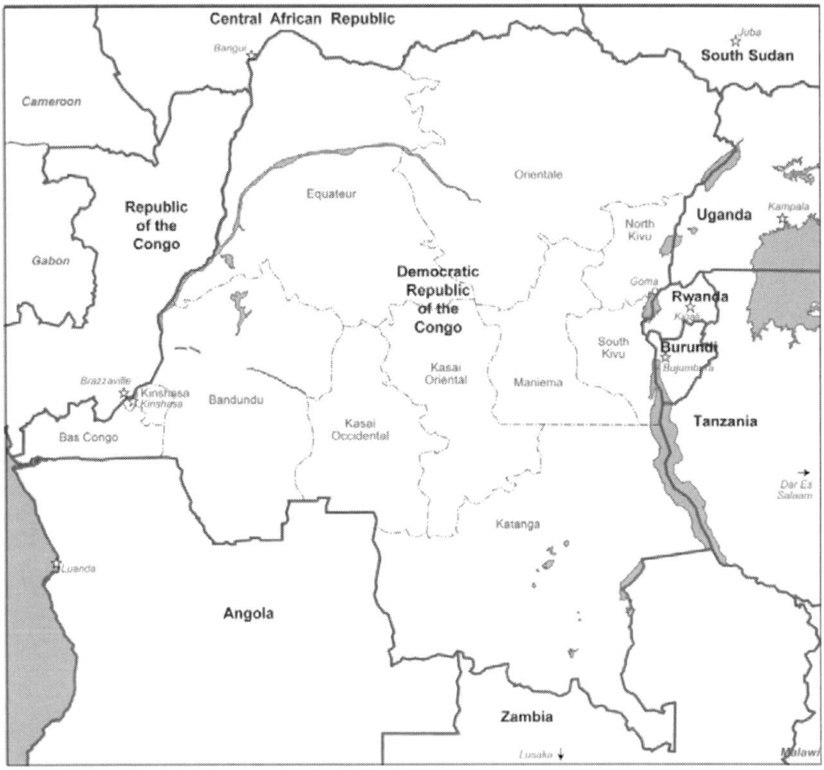

Source: GAO based on UN map.| GAO-14-575

Figure 1. Map of the Democratic Republic of the Congo with Provinces and Adjoining Countries.

U.S. Government Response to Conflict in the DRC

Congress has focused on issues related to the DRC for almost a decade. In 2006, Congress passed the Democratic Republic of Congo Relief, Security, and Democracy Promotion Act of 2006,[15] stating that U.S policy is to engage with governments working for peace and security throughout the DRC and hold accountable individuals, entities, and countries working to destabilize the government. In July 2010, Congress passed the Dodd-Frank Act, which included several provisions in section 1502 of the Act concerning conflict minerals in the DRC and adjoining countries. The Act directs State, USAID, SEC, and Commerce to take steps on matters related to the implementation of those provisions (see text box).

In addition, in July 2013, the United States appointed the current Special Envoy for the Great Lakes Region and the DRC, whose office develops and leads the implementation of U.S. regional policy on cross-border security, political, economic and social issues. The Special Envoy leads U.S. efforts to support the implementation of the Peace, Security, and Cooperation Framework Agreement, including the development and implementation of a comprehensive strategy to stop human suffering and violence in the region, by promoting political, economic, and social reconciliation.

International Response to Conflict in the DRC

In 1999, the UN Security Council authorized peacekeeping operations in the DRC, known as the UN Organization Mission in Democratic Republic of the Congo (MONUC).[16] MONUC's mission included achieving a ceasefire and protecting civilians and other nonmilitary personnel from threats of physical violence. In 2010, MONUC was replaced by the UN Organization Stabilization Mission in the Democratic Republic of the Congo (MONUSCO), whose priorities also include protecting civilians and stabilizing the country.

The international community has also responded to the conflict in the DRC and adjoining countries by appointing special envoys to the region. For example, in March 2013, the UN appointed a Special Envoy of the Secretary-General for the Great Lakes Region to support the implementation of the 11-nation Peace, Security and Cooperation Framework for the Democratic Republic of the Congo and the Region.[17] In June 2013, the UN also appointed the current Special Representative of the Secretary-General for the Democratic Republic of the Congo and Head of MONUSCO. Both the European Union and African Union also have appointed representatives for the Great Lakes Region. To address imminent threats to peace and security, the UN Security Council also authorized the deployment of a force intervention brigade within the current peacekeeping operations in the DRC on March 28, 2013. The objectives of the new force based in North Kivu province are to neutralize armed groups, reduce the threat they pose to state authority and civilian security, and make space for stabilization activities.[18]

PROVISIONS IN THE DODD-FRANK ACT RELATED TO CONFLICT MINERALS IN THE DRC AND ADJOINING COUNTRIES

- Section 1502(a) states that "it is the sense of the Congress that the exploitation and trade of conflict minerals originating in the Democratic Republic of the Congo is helping to finance conflict characterized by extreme levels of violence in the eastern Democratic Republic of the Congo, particularly sexual- and gender-based violence, and contributing to an emergency humanitarian situation therein, warranting the provisions of section 13(p) of the Securities Exchange Act of 1934, as added by subsection (b)."
- Section 1502(b) requires SEC, in consultation with State, to promulgate disclosure and reporting regulations regarding the use of conflict minerals from DRC and adjoining countries.
- Section 1502(c) requires State and USAID to develop, among other things, a strategy to address the linkages among human rights abuses, armed groups, the mining of conflict minerals, and commercial products.
- Section 1502(d) requires that Commerce report, among other things, a listing of all known conflict minerals processing facilities worldwide.

Source: GAO analysis. | GAO-14-575

In addition, the European Union (EU) is exploring possible legislation related to conflict minerals and responsible sourcing.[19] According to a European Commission release, in March 2014, the EU proposed a draft regulation setting up an EU system of self-certification for importers of tin, tantalum, tungsten, and gold for imports into the EU. The draft regulation indicated that the self-certification would align with the Organization for Economic Cooperation and Development's (OECD) "OECD Due Diligence Guidance for Responsible Supply Chains of Minerals from Conflict-Affected and High-Risk Areas," which includes a five-step framework for risk-based due diligence in the supply chain.[20] According to the release, the regulation gives EU importers an opportunity to deepen ongoing efforts to ensure clean supply chains when trading legitimately with operators in conflict-affected countries.

Conflict Minerals Supply Chains

Supply chains for companies using tin, tantalum, tungsten, and gold generally begin at the mine site, where ore is extracted from the ground with mechanized or artisanal mining techniques.[21] However, these supply chains can be complex and vary considerably, according to some industry association and company representatives. For example, as figure 2 shows, in the "upstream" segment of the supply chain—that is, from mine to smelter—ore may be purchased by a local processor or trader and then by an exporter, who ships it to a smelter for refinement; in other cases, the ore may be sold directly to an exporter. The "downstream" segments of conflict mineral supply chains—that is, from smelter to manufacturer—may vary as well, depending in part on the type of mineral. Figure 2 provides a simplified depiction of the supply chain for the four conflict minerals.

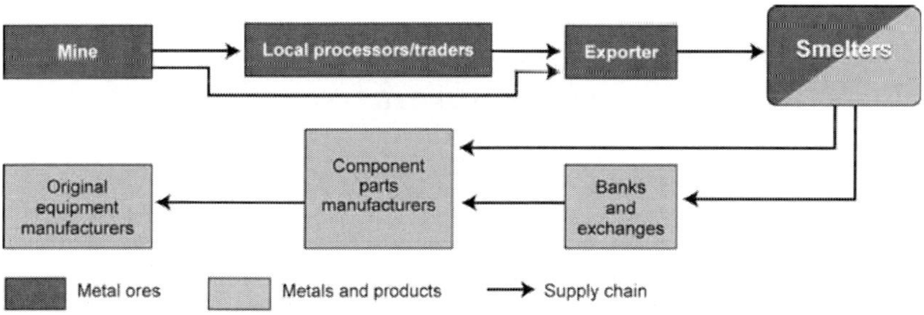

Source: GAO analysis. | GAO-14-575

Figure 2. Simplified Conflict Minerals Supply Chain.

USES OF CONFLICT MINERALS

Various industries, particularly manufacturing industries, use the four conflict minerals in a wide variety of products. For example, tin is used to solder metal pieces and is also found in food packaging, in steel coatings on automobile parts, and in some plastics.

Most tantalum is used to manufacture tantalum capacitors, which enable energy storage in electronic products such as cell phones and computers as well as used to produce alloy additives, which can be found in turbines in jet engines. Tungsten is used in automobile manufacturing, drill bits and cutting tools, and other industrial manufacturing tools and is the primary component of filaments in light bulbs. Gold is used as reserves and in jewelry and is also used by the electronics industry.

Source: GAO-12-763. | GAO-14-575.

Smelters and refiners are considered the "choke points" in the supply chain, since a limited number of smelters and refiners process conflict minerals worldwide and the origin of the minerals after processing can be difficult to verify. Smelters primarily provide high-purity tin, tantalum, and tungsten directly to component parts manufacturers, although some sell high-purity metals through traders or exchanges. Gold refiners typically sell high-purity gold to banks for use as a store of value or to international exchanges where gold is bought and sold,[22] although some refiners sell gold directly to manufacturers; banks and traders may also sell gold to manufacturers, including jewelry and component parts manufacturers. Component parts manufacturers use the refined tin, tantalum, tungsten, or gold to construct individual parts—such as capacitors, engine parts, or clasps for necklaces—that they sell to original equipment manufacturers. The original equipment manufacturers complete the final assembly of a product and sell the final product to the consumer.

Responsible Sourcing Initiatives

Various stakeholders—including governments, industry associations, international organizations, and international and local NGOs working in the Great Lakes Region—operate or support initiatives to promote and exercise responsible sourcing of conflict minerals. Stakeholder-developed initiatives—which include the development of guidance documents, audit protocols, and sourcing practices—support efforts by companies reporting to SEC under the rule to (1) conduct due diligence of their conflict minerals supply chain, (2) identify the source of conflict minerals within their supply chain, and (3) responsibly source conflict minerals. The initiatives can be divided into two categories: global or in-region.[23]

GLOBAL AND IN-REGION SOURCING INITIATIVES

Global sourcing initiatives may minimize the risk that minerals that have been exploited by illegal armed groups will enter the supply chain and may also support companies' efforts to identify the source of the conflict minerals across the supply chain around the world. In-region sourcing initiatives may support responsible sourcing of conflict minerals from Central Africa and the identification of specific mines of origin for those minerals. Such initiatives in DRC and adjoining countries focus on tracing minerals from the mine to the mineral smelter or refiner by supporting a bagging and tagging program or some type of traceability scheme.

Source: GAO | GAO-14-575

Most responsible sourcing initiatives follow OECD's due-diligence guidance.[24] The guidance is for use by any company potentially sourcing minerals or metals from conflict-affected and high-risk areas and, according to OECD, is one of the only international frameworks available to help companies meet their due diligence reporting requirements.

U.S. AGENCIES HAVE TAKEN ACTIONS RELATED TO RESPONSIBLE SOURCING OF CONFLICT MINERALS, BUT COMMERCE HAS NOT YET FULFILLED ITS REQUIREMENTS

Since the Act was passed in 2010, State, USAID, SEC, and Commerce have undertaken activities related to implementation of the Act's conflict minerals provisions, including activities related to responsible sourcing of such minerals from the DRC and adjoining countries. As required by the Act, State and USAID developed a strategy in 2011 aimed at addressing the linkages between human rights abuses, armed groups, the mining of conflict minerals, and commercial products and are implementing various objectives of the strategy. State also produced a map of mineral-rich areas under control of armed groups in the DRC. SEC issued its required conflict minerals rule in 2012. As of May 2014, Commerce had taken steps toward producing a list of all conflict minerals processing facilities worldwide, which the Act required by January 2013, but had not completed the task. Moreover, Commerce had not developed a plan of action with associated timeframes for how and when it expects to complete this effort and report to Congress. Standard practices in program and project management and execution include, among other things, developing a plan to execute specific projects needed to obtain defined results within a specific time frame.[25] State, USAID, SEC, and Commerce also have engaged in activities involving stakeholder partnerships and outreach and have provided technical assistance to other governments related to activities focused on the responsible sourcing of conflict minerals.

State, USAID, and SEC Have Addressed the Act's Requirements, and Commerce Is Working toward Fulfilling Its Requirements

The Act directed State, USAID, SEC, and Commerce to undertake various activities to implement its provisions related to conflict minerals. Since the Act was passed in 2010, the agencies have taken the following actions.

State and USAID Developed a Strategy, and State Produced a Conflict Minerals Map

Responding to the Act, State and USAID developed a strategy in 2011 to address the linkages among human rights abuses, armed groups, the mining of conflict minerals, and commercial products. The Act required State and USAID to submit, by January 2011,[26] a strategy to address the linkages between human rights abuses, armed groups, mining of conflict minerals, and commercial products. The strategy document that State and USAID submitted to Congress in 2011 lists five objectives: (1) promote an appropriate role of security forces, (2) enhance civilian regulation of minerals trade in the DRC, (3) protect artisanal miners and local communities, (4) strengthen regional and international efforts, and (5) promote due diligence and responsible trade through public outreach. The strategy includes activities corresponding with each of these objectives—for example, building the capacity of civilian mining authorities in the DRC to certify mine sites, supporting the implementation and coordination of certification and traceability schemes, building the capacity of the ICGLR related to mineral audit mechanisms, and engaging with industries and civil society groups regarding supply chain due diligence efforts. State and USAID officials indicated that they have been implementing objectives of the strategy over the past few years. According to the U.S. Special Envoy for the Great Lakes Region and the DRC, the strategy remains relevant and accurate and State has used it in conjunction with other U.S. government agencies as a roadmap for efforts to help break the link between armed groups and conflict minerals.

In addition, in 2011, State developed a map of mineral-rich zones and areas under control of armed groups in the DRC and has subsequently published several updated maps, as required by the Act.[27] The maps are focused on the exploitation of tin, tantalum, tungsten, and gold in the provinces of North and South Kivu and parts of Orientale, Maniema, and Katanga provinces. According to State, the most current map, which State published in February 2014, was based on data from surveys conducted in 2013 by the International Peace Information Service (IPIS)—an NGO— and on information from consultations with the DRC government, the UN Group of Experts, and MONUSCO (see app. II for the 2014 map).[28] State reported that lack of complete or fully verifiable data makes it difficult to confirm the location of many mine sites, to establish which mine sites are active at any given time, and to comprehensively verify reports of armed groups or other entities that are either present at mines or have access to revenue streams emanating from them. State officials indicated that in the future the map may become digital rather than paper based.

SEC Issued Its Conflict Minerals Rule in 2012

As we previously reported,[29] SEC issued its conflict minerals rule in August 2012.[30] The Act required that SEC promulgate, by April 2011, disclosure and reporting regulations regarding the use of conflict minerals from the DRC and adjoining countries.[31]

SEC issued a "frequently asked questions" (FAQ) document in May 2013 to address questions by companies that will have to report to SEC under the conflict minerals rule. SEC officials indicated that these FAQs included questions posed most often by companies regarding interpretation of the rule. In April 2014, SEC issued additional FAQs addressing questions that mostly pertained to the independent private sector audit of companies' conflict minerals disclosure reports. According to an SEC official, these FAQs were based on interpretive questions asked by SEC-reporting companies and the audit community.

In January 2014, SEC made "Form SD," a specialized disclosure form for reporting compliance with the conflict minerals rule, available for electronic filing. The form, originally published with the conflict minerals rule, provides general instructions to SEC-reporting companies for filing the conflict minerals disclosure and specifies the information that their conflict minerals reports must include. SEC-reporting companies were required to file under the rule for the first time by June 2, 2014, and annually thereafter on May 31.[32] According to SEC officials, based on preliminary feedback they received, they anticipated that most SEC-reporting companies subject to the rule would be unable to determine whether or not their products qualified as "DRC conflict-free."[33]

Commerce Has Taken Steps but Has Not Provided a List of Conflict Minerals Processing Facilities to Congress

More than a year after the deadline required by the Act, Commerce has not yet fulfilled its mandate under section 1502 of the Act. Section 1502 directed Commerce to report, among other things, a list of all known conflict minerals processing facilities worldwide to appropriate congressional committees annually starting no later than 30 months after the Act's enactment—that is, by January 2013.[34] As of May 2014, Commerce had not developed such a list or developed a plan of action, with associated time frames, for completing this requirement and reporting it to Congress. Standard practices in program and project management include developing, among other things, a program plan to execute specific projects needed to obtain defined programmatic results within a specific time frame. In January 2014, Commerce officials told us that they had identified entities that they hoped would help them identify publicly available information about conflict minerals and identify stakeholders who are knowledgeable about conflict minerals issues. Specifically, Commerce officials indicated that they had assembled a proposed and internal outreach plan, which includes meeting with stakeholders to discuss how these organizations have gathered information on conflict mineral smelters and identifies other efforts that Commerce can explore to develop the list of conflict minerals processing facilities. Commerce officials also indicated that they anticipated a 3- to 4-month time frame for the proposed outreach efforts to talk to stakeholders. In May 2014, Commerce officials stated that they had completed discussions with the majority of the stakeholders identified in the outreach plan and have developed several preliminary lists of conflict minerals processing facilities, based on information they obtained from the stakeholders. However, Commerce officials stated that they did not have a timeframe for completing the final list for Congress.

Commerce officials said that they had encountered some challenges associated with gathering data on conflict minerals to help inform their outreach plan and required reporting. For example, according to the officials, conflict minerals and mining operations are difficult to track; because the equipment used to process conflict minerals can be moved easily, such operations can emerge in different locations. In addition, Commerce officials mentioned that some conflict minerals data may be inaccessible to the U.S. government because a large number of conflict mineral smelters are in China.[35] Having an action plan with associated timeframes could better position Commerce to report on the status of its efforts to compile a list of conflict minerals processing facilities worldwide and to hold its personnel accountable for completing its related activities.

Some stakeholders that we contacted, including government and industry officials and representatives of the UN Group of Experts and an NGO, indicated that a comprehensive list of conflict minerals smelters and refiners—considered the "choke point" of the supply chain—would be very useful in the effort to ensure responsible sourcing of minerals in the DRC and adjoining countries. According to these stakeholders, such a list would enable companies that are subject to the SEC rule to maintain transparency regarding their supply chains, particularly in their communications with smelters, and would also provide companies the information they need for their SEC-required conflict minerals disclosure reports.

U.S. Agencies Have Engaged in Partnerships and Outreach and Provided Technical Assistance to Stakeholders

U.S. government agencies have engaged in a variety of activities that involve partnerships and coordination with other stakeholders or outreach to stakeholders, and some agencies have provided technical assistance to stakeholders regarding responsible sourcing of conflict minerals. Some agencies' activities contribute to global and in-region responsible sourcing initiatives and some of the activities address the implementation of objectives outlined in the strategy to address the linkages between human rights abuses, armed groups, and the mining of conflict minerals.

U.S. Agency Partnerships and Coordination Efforts

Some U.S. agencies have partnered and coordinated with other stakeholders—other government agencies, industry, and civil society—regarding issues related to responsible sourcing of conflict minerals. For example:

- State and USAID, both in headquarters and posts or missions overseas, and other U.S. agencies coordinate with one another on weekly or biweekly conference calls to discuss the progress of responsible sourcing efforts, provide updates on recent events, and collaborate on future events, according to State and USAID officials.
- USAID works in a collaborative and coordinated manner with State in Washington and regionally, using the 2011 U.S. strategy as a framework for the coordination, according to USAID officials. The officials indicated that funding also has been coordinated between the two agencies across the five objectives of the strategy and totals over $25 million, as of 2013.
- State and USAID coordinate with other stakeholders through the Public-Private Alliance for Responsible Minerals Trade (PPA) to fund and support organizations working on responsible sourcing efforts. Both State and USAID are on the PPA's Governance Committee, which consists of participants from foreign governments, industry, and civil society.[36]
- USAID has partnered with the International Organization for Migration to help enhance civilian control of the DRC's mineral trade through infrastructure improvements and institutional reforms, according to agency officials. The officials reported that with USAID funding, the organization will also establish pilot certification and traceability systems in and around the mineral trading centers and other areas of South and North Kivu.[37]
- USAID has partnered and coordinated with stakeholders in the DRC, according to agency officials. For example, the officials said that USAID is coordinating with the DRC government regarding various aspects of minerals trade, is involved in the multi-stakeholder Mining Thematic Group in the DRC, and facilitates the Eastern Congo Mining Coordination Team.
- On a multilateral level, both USAID and State participate in the OECD Responsible Sourcing Stakeholder Forums held every 6 months, according to USAID officials. This forum, coordinated by the ICGLR, OECD, and the UN Group of Experts, is a platform for governments, the private sector, international organizations, and civil society to share experiences with implementation of supply chain due diligence for responsible sourcing of minerals from conflict-affected and high-risk areas. Both State and USAID officials have participated at times as facilitators of these forums.

- The current U.S. Special Envoy for the Great Lakes Region and the DRC has collaborated with other stakeholders, such as the UN Special Envoy for the Great Lakes, the African Union, and other multilateral and bilateral partners, to strengthen international coordination mechanisms on the crisis in the Great Lakes, according to State. These efforts have taken place under the Peace, Security, and Cooperation Framework Agreement for the DRC and the Region.

U.S. Agency Outreach Efforts

A couple of U.S. government agencies indicated that they have conducted outreach to various stakeholders to promote responsible sourcing of conflict minerals and to obtain information about conflict mineral sourcing and supply chains. For example, State and Commerce officials reported the following.

- State officials told us that State has engaged with foreign governments and industry associations regarding the Dodd-Frank Act requirements. According to these officials, State's efforts have included sending letters about section 1502 of the Act to foreign governments that are prominent in the conflict mineral supply chain as well as encouraging these governments and companies in those countries to support the aim of the legislation. In a November 2013 briefing, the Deputy Assistant Secretary of State for Counter Threat Finance and Sanctions reported that he had travelled to Asia and Europe to talk with representatives from smelters and governments about responsible sourcing initiatives and encourage participation in such initiatives. State officials also indicated that they have facilitated outreach efforts for industry associations, such as the Conflict-Free Smelter Initiative (CFSI) and others, to help them secure meetings in Asian countries to discuss conflict free mining and smelting. According to the State officials, during the outreach some members from industry and industry associations expressed interest in talking to Commerce about smelters and support for responsible sourcing, according to State officials.
- Commerce officials stated that their proposed outreach plan identified entities that could enable them to develop the list of conflict mineral smelters and refiners required by the Act. According to these officials, they have conducted outreach to these entities, including government agencies, industry associations, international organizations, and NGOs.

U.S. Agency Technical Assistance

Some U.S. agencies have provided technical assistance related to responsible sourcing of conflict minerals to various stakeholders. According to agency officials, these stakeholders have consisted primarily of other governments, particularly in the Great Lakes Region. For example:

- State officials said that they had shared experiences and challenges related to implementing the Act with officials from the EU who were working on proposed conflict minerals legislation.
- SEC officials stated that they had discussed with EU officials issues that SEC considered when drafting the conflict minerals rule as well as questions about the rule that SEC received from industry.
- USAID officials stated that they had been working with the ICGLR in providing technical assistance on conflict minerals programs, particularly through the steering committee for ICGLR's Regional Initiative against the Illegal Exploitation of Natural Resources.[38] Specifically, according to USAID officials, USAID has been implementing a multiyear institutional capacity program in support of the ICGLR to build the overall strength of the Executive Secretariat as well as the ICGLR's regional initiative. USAID officials said that the agency will soon begin implementing activities to support a third-party supply chain audit mechanism and an independent conflict minerals supply chain auditor.

Stakeholder Initiatives Are Expanding and Provide Some Information on Sourcing of Conflict Minerals

Since we reported in July 2013,[39] stakeholders have expanded existing initiatives and added new initiatives focused on responsible sourcing of conflict minerals in the DRC and adjoining countries, to include new mine sites, countries, and smelters. Some of these initiatives have yielded publically available information, including data on production of conflict-free minerals and export data, as well as reports on the progress and results of the initiatives. However, this information is limited in scope and thus may not provide a comprehensive description of the sourcing of conflict minerals from the DRC and adjoining countries.

Stakeholders Have Expanded Existing Initiatives, and New Initiatives Are Underway

Stakeholders have recently expanded, or made plans to expand, a number of existing global and in-region responsible sourcing initiatives; and two new initiatives are underway. Figure 3 shows the starting dates for existing, expanding, and new responsible sourcing initiatives. According to some stakeholders we interviewed, improvement in security in eastern DRC and industries' growing awareness of responsible sourcing and the Act's requirements may account for the expansion of responsible sourcing initiative.

Some Existing Global Initiatives Are Expanding

The following are examples of global responsible-sourcing initiatives that stakeholders have expanded since we reported in July 2013.[40]

Conflict-Free Sourcing Initiative

The Conflict-Free Sourcing Initiative (CFSI) has expanded in several aspects related to responsible sourcing.[41] First, CFSI's Conflict-Free Smelter Program has expanded the number of smelters it has certified as conflict free. The program is a voluntary one in which smelters undergo an independent third-party audit, in accordance with OECD's due diligence guidelines, to verify the origin of minerals processed at their facilities. The number of smelters that the program has certified as conflict-free has expanded from 26 smelters in summer 2013 to 85 smelters as of April 25, 2014 (see table 1). An additional 25 smelters are in the process of being certified, bringing the total number of smelters involved in the program to 110.[42] As of January 2014, the Conflict-Free Smelter Program has expanded to include smelters for tungsten in addition to the other three conflict minerals.

Second, according to CFSI representatives, through outreach to industry, CFSI has expanded its collaboration with companies involved with the conflict minerals supply chain. CFSI's outreach includes twice-yearly workshops on conflict minerals issues that are open to all participants. According to CFSI, outreach such as these workshops bring together hundreds of representatives from industry, government, and civil society for updates, in-depth discussions, and guidance on best practices for responsible mineral sourcing.[43] CFSI officials stated that such outreach recently resulted in collaboration with the tungsten industry, which led to certification of the first conflict-free tungsten smelter in 2014.

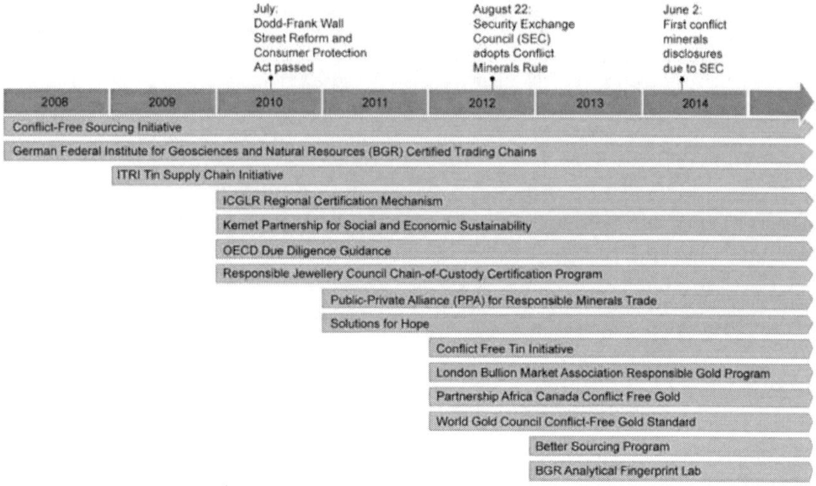

Source: GAO analysis. | GAO-14-575

Figure 3. Timeline of the Starting Dates of Stakeholder-led Initiatives from 2008 to June 2014.

Table 1. Number of Smelters and Refiners in the Conflict-Free Smelter Program as of April 25, 2014

Mineral	Number of certified smelters/refiners	Number of smelters/ refiners working toward certification	Total
Tantalum	28	1	**29**
Tin	13	14	**27**
Tungsten	1	8	**9**
Gold	43	2	**45**
Total	**85**	**25**	**110**

Source: Conflict-Free Sourcing Initiative data, GAO (analysis). | GAO-14-575

Third, in 2014, CFSI began offering its members information about the SEC-required "reasonable country of origin" data for conflict minerals, providing the most detailed information currently available about the source of conflict minerals for smelting and refining facilities that are validated through the Conflict-Free Smelter Program. According to CFSI, this information may be useful to companies as they prepare the conflict minerals disclosure reports required by the SEC rule and demonstrate conformance with the OECD due diligence guidelines.

Responsible Gold Program

In January 2012, the London Bullion Market Association (LBMA), which represents the global market for gold and silver, finalized and published its Responsible Gold Guidance to ensure that the gold refiners it accredits purchase only conflict-free gold.[44] The refiners accredited by LBMA are required to complete an annual third-party audit to verify their compliance with the LBMA guidance, according to an LBMA official. As of March 2014, of the 67 gold refiners that LBMA oversees, more than three-quarters had successfully submitted their audits and received the Responsible Gold Certificate, according to the official. The representative stated that if a refiner does not submit a third-party audit by the end of 2014, the refiner will be removed from LBMA's list of accredited refiners. An LBMA official said that the association also collaborates with other responsible sourcing stakeholders and global gold exchanges and works closely with OECD. For example, the official said that, working through OECD, LBMA has met with Chinese industry representatives to clarify the purpose and benefits of conducting due diligence audits of their refiners.

Responsible Jewellery Council Chain-of-Custody Certification Program

The Responsible Jewellery Council—a diamond and precious metals industry association—launched a chain-of-custody certification program in March 2012 to help its member companies identify and track conflict-free gold throughout their supply chains.[45] The program's requirements, which are aligned with the OECD Due Diligence Guidance for gold, include a third-party audit of each certified entity to ensure that its gold is conflict-free, according to the Responsible Jewellery Council. According to an official with the Responsible Jewellery Council, this certification can support companies' compliance with the Dodd Frank Act. As of April 2014, nine entities had been validated under the council's certification program and more entities were in the process of being certified, according to the official.[46]

Some In-Region Initiatives Have Also Expanded

The following are examples of in-region responsible sourcing initiatives that stakeholders have expanded, or made plans to expand, since we reported in July 2013.

ITRI Tin Supply Chain Initiative

The ITRI Tin Supply Chain Initiative (iTSCi) recently announced that it is expanding its in-region operations. The initiative works with "upstream" entities (i.e., companies involved in the conflict minerals supply chain from mine to smelter) in instituting the actions, structures, and processes necessary to conform with the OECD Due Diligence Guidance and helps relevant U.S. companies report on their due diligence efforts to the SEC as required by the Dodd-Frank Act.[47] The assistance that iTSCi provides includes a system to trace bags of minerals from the mines to the exporter, due diligence audits of iTSCi's member companies, and assessments of the political and security situations, which have been conducted at various mine sites in the DRC and Rwanda.

In February 2014, iTSCi announced that it was expanding its traceability program into a remote area in the northern region of the Maniema province of the DRC, and into the North Kivu province of the DRC. According to iTSCi, improved security in North Kivu, which has a history of armed conflict, accounts in part for the expansion into the province. An iTSCi official stated that the initiative is currently looking at options for extending into South Kivu. Additionally, in April 2014, iTSCi announced that the program had started operations in Burundi. According to iTSCi, there is presently little evidence of activity by nongovernment armed groups in Burundi, since there have been no reports that armed groups are controlling mine sites or transportation routes, extorting money or minerals, or illegally taxing the trade of minerals. iTSCi further reported that it may extend the program to Uganda and eventually to the entire Great Lakes Region. Also, an iTSCi official stated that the initiative had successfully piloted technology in Rwanda to collect and manage data on conflict minerals electronically, which would replace the current paper-based system and increase efficiency of data collection.

Solutions for Hope

Launched by Motorola Solutions and AVX in 2011, the Solutions for Hope tantalum program is a "closed-pipeline" initiative that traces the flow of tantalum from the mine to the end-use company.[48] In June 2013, Solutions for Hope reported that it had completed six shipments of tantalum, totaling more than 145 metric tons, from the Katanga province in the DRC. According to officials, in part because of improved security in the province, the initiative started sourcing tantalum from North Kivu in March 2014. Officials also noted that Solutions for Hope is exploring a closed pipeline system for gold in the DRC.

Conflict-Free Tin Initiative

The Conflict-Free Tin Initiative (CFTI), a multistakeholder effort supported by the Netherlands government, is a closed-pipeline initiative, similar to Solutions for Hope, started in October 2012 for sourcing tin from the South Kivu province of the DRC. According to CFTI, the initiative has expanded its mining operation to Maniema, a province bordering South Kivu, which is less prone to conflict and the government is reinvesting tax income into the mining communities. According to USAID, a CFTI stakeholder, from October 2012 to December 2013, the initiative generated a total export value of more than $3 million.

Regional Certification Mechanism

In 2010, the International Conference on the Great Lakes Region (ICGLR) began working with an NGO to develop a regional certification mechanism to ensure that conflict minerals are fully traceable. ICGLR's Regional Certification Mechanism (RCM) enables member countries and their mining companies to demonstrate where and under what conditions minerals were produced, allowing member governments to issue ICGLR regional certificates for those mineral shipments that are in compliance with the standards of the mechanism. The ICGLR issued its first certificate in November 2013 to a mine in Rwanda. According to an ICGLR official, the DRC launched its certificate program but had not yet issued any certificates as of November 2013. He added that Tanzania and Burundi may be able to issue certificates by the end of 2014. The ICGLR official noted several challenges in instituting the RCM in the DRC and the region. For example, he cited that it is logistically difficult to catalogue all mines in each country. In addition, the official noted that training local officials to use the RCM software is difficult and time consuming. The official added that it takes member countries 1 year to prepare for all components associated with launching the RCM.

New Initiatives Have Launched in the Region Since 2013

In the past year, one existing stakeholder has launched a new in-region responsible sourcing initiative and a new stakeholder has established an initiative.

Analytical Fingerprint Project

The German government's Federal Institute for Geosciences and Natural Resources (BGR), an existing stakeholder, launched a new initiative in the past year. According to a representative, BGR's primary role is to support the region's governments and to build government capacity. Since 2013, BGR has initiated the Analytical Fingerprint Project to allow for independent verification of the origin of the conflict mineral by comparing the composition of tantalum, tin, and tungsten concentrate samples of a known origin with unknown samples, similar to a DNA test. According to a BGR official, the project has three types of units—sample preparation labs, high tech labs, and a management unit. To date, BGR has established sample preparation labs, located in Rwanda since 2013 and in the DRC since 2014, where mineral ore samples are prepared for analysis. A third sample preparation lab is under construction in Burundi. BGR is in negotiations to establish a high tech lab in Tanzania, which receives and analyzes the samples from the preparation labs. Additionally, in 2013, according to a BGR official, BGR established a project management unit at the ICGLR headquarters in Burundi, which evaluates the raw data and produces the analytical fingerprints.

Better Sourcing Program Responsible-Sourcing Pilot

The Better Sourcing Program (BSP), a private company that offers an independently audited due diligence assurance program to enable companies to source tantalum, tin, tungsten, and gold from the region, is a new stakeholder in the region in 2013. The Better Sourcing Program established a pilot program that covers a tantalum supply chain originating from the Republic of the Congo (also known as Congo-Brazzaville), which is the first responsible-sourcing initiative in that country. Better Sourcing Program officials stated that they chose to pilot the initiative in the Republic of the Congo because no other scheme to support producers existed there, because the country is relatively conflict free, and because the government has been cooperative.

Stakeholders Face Operational Challenges in Expanding or Launching In-Region Initiatives

Some stakeholders that we interviewed noted various challenges to expanding or launching responsible-sourcing efforts in the DRC and adjoining countries. For example:

- **Lack of infrastructure.** Some stakeholders reported that inadequate infrastructure in the DRC and adjoining countries affects their ability to operate in the region and expand initiatives to new areas. Representatives from Solutions for Hope stated that infrastructure in the DRC cannot support a large-scale smelter. According to these representatives, the power supply in the DRC can be inconsistent and, because smelting facilities require a large, consistent power supply to function properly, all tin, tantalum, and tungsten currently are exported from the DRC and Rwanda for smelting.[49]

- **Lack of government support.** Some stakeholders reported some operational challenges related to the region's national and provincial governments. For example, stakeholders involved in the Conflict Free Tin Initiative stated that sales of tin from their mine in South Kivu were halted for nearly 2 months in 2013 after the provincial government of the region imposed harsh taxes on the minerals mined there. Additionally, Solutions for Hope officials reported that the DRC's current tax structure is not conducive to legitimizing gold and that expanding the initiative to include gold is therefore difficult.
- **Lack of buyers for conflict minerals from conflict zones.** Several stakeholders and agency officials reported that some companies are reluctant to buy minerals produced in the DRC and adjoining countries because of the high cost of the due diligence required by the Dodd-Frank Act and the perceived reputational risk. For example, according to one industry official, a major challenge to responsible sourcing in the region is that the cost of complying with the SEC rule makes it difficult for SEC-reporting companies to compete in the global market against companies that are not required to perform costly due diligence. In addition, an official from the Responsible Jewellery Council noted that gold is mined in many locations around the world and that production costs must always be taken into account. She stated that because the cost of due diligence for gold is usually proportional to risk, mining and sourcing responsibly in the Great Lakes Region could become more expensive than in other, lower-risk areas and that this represents a challenge to responsible sourcing efforts in the region.

Some Information about the Conflict Minerals Trade Is Publicly Available

Some stakeholders and governments in the region provide publicly available information related to in-region mining of conflict minerals and responsible sourcing initiatives. According to industry officials, the amount of publicly available data reported by responsible sourcing initiatives has increased over the past year. We found that iTSCi publishes various reports on its public website as part of its due diligence system for its members. These reports provide production and export data for tin, tantalum, and tungsten, including amounts traced through the iTSCi program, from three provinces in the DRC and Rwanda, as well as the mineral sales in U.S. dollars (for more details, including the applicable quantitative data, see app. IV). Also on its website, iTSCi publishes third-party audits of member companies, which assess the extent to which the companies have implemented the OECD Due Diligence Guidance and evaluate the companies' adherence to iTSCi's traceability and due diligence procedure. Additionally, iTSCi publishes governance assessments covering a range of topics. Examples include the security and political situation in areas without an iTSCi presence and the risks and performance, relative to the OECD Due Diligence Guidance and the SEC rule, of stakeholders that are part of, or play a role in monitoring, the conflict minerals supply chain.[50]

A comparison of production data from the conflict-free sourcing initiatives in the context of each country's or the region's total production or exports of tin, tantalum, tungsten, or gold is not feasible, because no reliable, comprehensive data on the production or export of conflict minerals for the countries or region are available. However, some quantitative government data on the production and exports of conflict minerals from the DRC and adjoining countries are available (see app. V). For example, the DRC government has published data on production and exports of all four conflict minerals and the Rwandan government has published export and value data for tin, tantalum, and tungsten. In addition, the International Trade Centre, a joint agency of the World Trade Organization and the UN, collects export data from governments that provide some context for the amount of conflict minerals declared as exported from the region, although these data do not identify minerals from conflict-free mines (see app. VI).

Some stakeholders indicated that the ICGLR may, at some point, be able to provide production and export data of conflict minerals from its member states, mostly Dodd-Frank-affected countries. This information could increase transparency at the individual country level. ICGLR requires its member states to implement a chain-of-custody tracking system for conflict minerals and to transmit data on mineral flows (i.e., quantities and destinations) at regular intervals to be incorporated in the ICGLR Regional Mineral Tracking Database. According to an ICGLR official, this database is populated with three types of information: (1) a historical record of mine sites in each country, (2) types and quantities of minerals produced at each mine site, and (3) mineral flows from the mine sites. According to ICGLR documentation, the data are used to track, analyze, and reconcile regional mineral flows and will become publicly available to ensure ICGLR's credibility. In November 2013, the ICGLR official stated that four of the 12 member countries had submitted information for the regional database: Rwanda, Uganda, the DRC, and, to a lesser extent, Burundi.

Little Additional Information on Rate of Sexual Violence in Eastern DRC and Adjoining Countries since July 2013

Since we reported in July 2013, no new population-based surveys related to sexual violence in the DRC, Uganda, Rwanda, or Burundi have been published. However, population-based surveys are underway, or being planned, in three of those countries—the DRC, Burundi, and Rwanda. In addition, some new case file data on sexual violence are available for all four countries. However, as we reported in 2011, case file data on sexual violence are not suitable for estimating a rate of sexual violence.

Several Population-Based Surveys Are Underway or Planned

Although no new surveys related to sexual violence in the DRC, Uganda, Rwanda, or Burundi have been published since July 2013, population-based surveys in the DRC, Burundi, and Rwanda are underway or planned by ICF International.[51] According to ICF International, data collection for a Demographic and Health Survey (DHS), which is a type of population-based survey, for the DRC is complete, but data resulting from the survey are not expected until September 2014 or later. ICF International also said that fieldwork for a DHS in Rwanda is likely to start in September or October 2014. ICF International indicated that a DHS is planned to start in Burundi in 2014; however, data collection may be delayed by funding gaps and, as a result, the survey may not take place until 2015.

Figure 4 shows the anticipated timelines for the population-based surveys on sexual violence that are currently underway or planned in the DRC, Burundi, and Rwanda. It also shows the publication dates for eight population-based surveys that provided data on the rate of sexual violence in eastern

DRC, Rwanda, and Uganda that have been published since we started reporting on sexual violence in the region in 2011.

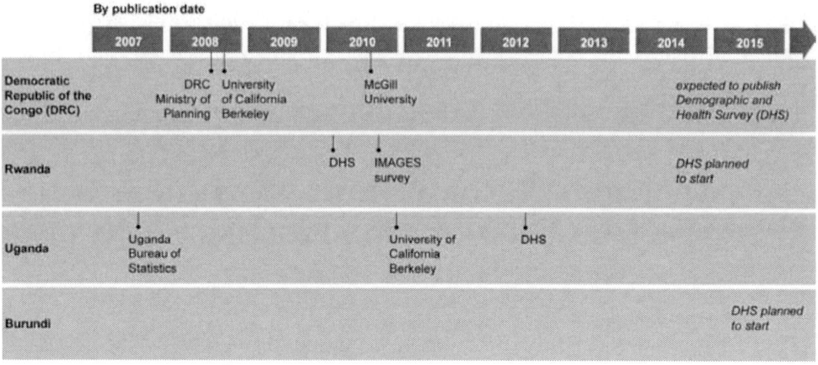

Source: GAO anbalysis.| GAO-14-575

Figure 4. Timeline of Population-Based Surveys Estimating the Rate of Sexual Violence in Eastern DRC, Rwanda, Uganda, and Burundi.

Some Additional Case File Data on Sexual Violence Have Become Available since GAO's 2013 Report

Since 2013, some U.S. and UN agencies as well as researchers and an NGO have provided additional case file data on instances of sexual violence in the DRC and adjoining countries. In February 2014, State submitted its annual country reports on human rights practices to Congress, which provided the following data pertaining to sexual violence in the DRC, Burundi, and Uganda.[52] For example,

- In the DRC, the government reported 18,729 cases of sexual violence in 2012.
- In Burundi, 3,781 cases of gender-based violence were reported in 2010, according to a report compiled from family development centers throughout the country.
- In Uganda, 530 cases of rape were registered in 2012, and 301 of the alleged rapists were tried and convicted.

In addition, some UN entities reported case file information. For example:

- MONUSCO reported that between October 1, 2013, and December 5, 2013, it recorded acts of sexual violence against at least 79 women and 28 girls in conflict-affected provinces in the DRC.[53]
- MONUSCO reported in March 2014 that sexual violence crimes continued to be committed by armed groups and that such crimes were allegedly committed against at least four women, 35 girls, and one man by elements of illegal armed forces and members of the Congolese military and police in January 2014.
- The United Nations Joint Human Rights Office reported in April 2014 that between the period of January 2010 and December 2013, it registered 3,635 victims of sexual violence throughout the DRC.[54] The report indicated that, for the reporting period, while more than half of the total alleged acts of sexual violence were committee by illegal armed groups, members of the Congolese national military committed less than half of the other alleged acts.

Moreover, the DRC government reported case file information. For example:

- A report published in June 2013 by the DRC Ministry of Gender, the Family, and Child, with support by the UN Population Fund, highlighted cases of sexual violence in seven provinces in the country: Bandundu, Bas Congo, Katanga, Kinshasa, North Kivu, Orientale, and South Kivu.[55] Data in the report indicate that 10,322 incidents of sexual and gender-based violence were reported for the seven provinces in 2011 and that the number increased to 15,654 incidents in 2012. While the data also indicate that most of these assaults were committed by people dressed in civilian clothes—85 percent in 2011 and 78 percent in 2012—incidents involving armed groups between 2011 and 2012 increased in South Kivu from 36 to 76 percent, respectively, and in North Kivu from 32 to 61 percent, respectively.

An NGO also published case file information. For example:

- Médecins sans Frontières reported in March 2014 that in 2012 it provided medical care to a total of 4,037 women, men, and children after incidents of sexual violence in different project locations in the DRC.[56] It reported treating, in a 5-week period in late 2012 through early 2013, 95 of those individuals in one camp for Internally Displaced People in North Kivu.

Several factors make case file data unsuitable for estimating rates of sexual violence. First, because case file data are not aggregated across various sources, and because the extent to which various reports overlap is unclear, it is difficult to obtain complete data, or a sense of magnitude, from case files. Second, in case file data as well as surveys, time frames, locales, and definitions of sexual violence may be inconsistent across data collection operations. Third, case file data are not based on a random sample and the results of analyzing these data are not generalizable.

CONCLUSION

The long-running humanitarian crisis in eastern DRC, one of the most volatile areas in Africa, continues to be a concern for the U.S. government and the international community. As we have previously reported, section 1502 of the Dodd-Frank Wall Street Reform and Consumer Protection Act, enacted in 2010, is part of the U.S. effort to address the perpetration of sexual violence and mass killings in the DRC by armed groups who profit from the exploitation and trading of conflict minerals. We have compiled stakeholder data on the production and trade of conflict minerals that demonstrates, to some extent, the degree of transparency related to the conflict minerals trade. The actions undertaken by U.S. agencies in response to the Act could facilitate SEC-reporting companies' compliance with the SEC rule, promulgated pursuant to the Act, as they conduct due diligence and prepare the required annual reports disclosing the use and the origin of conflict minerals in their products. However, because Commerce has not met the Act's requirement that it compile by January 2013 a list of smelters and refiners—considered the "choke point" of the conflict minerals supply chain—these companies lack a source of critical information about the conflict minerals supply chain.

Some stakeholders indicated that a comprehensive list of conflict minerals smelters and refiners could enable companies that are subject to the SEC rule to maintain transparency of the supply chain, and also provide companies the information they need for their SEC-required conflict minerals disclosure reports. Commerce cited several challenges that have hindered its providing the required list to Congress.

However, having an action plan with associated timeframes could better position Commerce to report on the status of its efforts to produce a final list and provide it to Congress and to hold its personnel accountable for completing related activities.

RECOMMENDATION FOR EXECUTIVE ACTION

To give Congress a sense of Commerce's efforts to produce a listing of all known conflict minerals processing facilities worldwide, as required by section 1502 of the Dodd-Frank Wall Street Reform and Consumer Protection Act, we recommend that the Secretary of Commerce provide to Congress a plan that outlines the steps, with associated timeframes, to develop and report the required information about smelters and refiners of conflict minerals worldwide.

AGENCY AND THIRD PARTY COMMENTS AND OUR EVALUATION

We provided a draft of this report to SEC, State, USAID, and Commerce for their review. Commerce provided written comments. SEC and State provided technical comments, which we incorporated as appropriate. We also provided relevant portions of the draft report to some industry associations and other stakeholders of conflict minerals initiatives from whom we had obtained information during our review; some of these stakeholders provided technical comments that we incorporated as appropriate.

In its written comments, Commerce noted in response to our recommendation that it had provided and briefed us on a detailed outreach action plan that set forth how it intended to assemble the list of conflict minerals processing facilities required by the Act. However, as we have noted in this report, the document that Commerce provided was an outreach plan, consisting of a list of stakeholders that Commerce intended to contact to obtain information on conflict minerals smelters and refiners.

The plan did not indicate a timeframe for completing and submitting to Congress the required listing of conflict minerals processing facilities worldwide.

Commerce concurred with our recommendation and noted that it will submit a listing of all known conflict minerals processing facilities worldwide to Congress by September 1, 2014.

Kimberly M. Gianopoulos
Acting Director, International Affairs and Trade

APPENDIX I. OBJECTIVES, SCOPE, AND METHODOLOGY

To determine the extent, if any, to which relevant U.S. agencies have engaged in activities related to responsible sourcing of conflict minerals, we interviewed officials who are cognizant of conflict minerals issues from the Departments of Commerce and State (State) and the United States Agency for International Development (USAID), and the Securities and Exchange Commission (SEC). We reviewed Section 1502 of the Dodd-Frank Wall Street Reform and Consumer Protection Act (Pub. L. No. 111-203) to identify the requirements for Commerce, State, SEC, and USAID related to implementation of section 1502. We also reviewed and analyzed reports and other documents from the agencies, such as the U.S. Strategy to Address the Linkages Between Human Rights Abuses, Armed Groups, Mining of Conflict Minerals, and Commercial Products; the conflict minerals rule; and maps of mineral-rich zones and areas under control of armed groups in the DRC. In addition, we reviewed and analyzed press releases, statements, plans, and guidance pertaining to conflict minerals and responsible sourcing that were issued by U.S. agencies. We also reviewed notes and agendas of responsible sourcing forums and other meetings attended by U.S. officials.

To analyze what is known about the status of, and any information provided by, initiatives focused on responsible sourcing of conflict minerals from the DRC and adjoining countries, we interviewed officials and reviewed and analyzed documents from State, USAID, and the United Nations Group of Experts on the Democratic Republic of the Congo (UNGoE); interviewed representatives and reviewed and analyzed guidance documents, reports, and presentations from foreign government, industry associations, multilateral organizations, companies, and nongovernmental organizations (NGOs). We selected these stakeholders based on their expertise on responsible sourcing issues, because they represented a range of perspectives on conflict minerals, and because we had established contacts with these entities on our last review.[57] In addition, some of the stakeholders we talked to have been working on the ground in the DRC. The stakeholders we spoke with constitute a nongeneralizable sample, and the information we gathered from them cannot be used to infer views of other stakeholders cognizant of conflict minerals issues. To determine which initiatives had expanded and which were new initiatives in the region, we interviewed U.S. agency officials and relevant stakeholders and reviewed documentation from initiatives in the region. This report covers initiatives on which we have previously reported and new initiatives since our 2013 report, as described by stakeholders we interviewed.[58] However, it is possible that the agency officials and stakeholders with whom we spoke may be unaware of other stakeholders and/or responsible sourcing initiatives active in the DRC and region. To determine the starting dates of the initiatives, we interviewed stakeholders and reviewed the websites of the various initiatives. To demonstrate what

information stakeholders have reported regarding responsible sourcing initiatives, we reviewed and analyzed information published on the websites associated with the various responsible sourcing initiatives. When reporting on information provided by the initiatives and stakeholders, we are referring to information such as reports published, amount of minerals mined per region/country, amount of minerals exported, and value of minerals produced and exported, including data covering 2012 through 2014. The information gathered cannot be generalized and cannot be used to infer views of other stakeholders cognizant of conflict minerals issues.

To demonstrate other sources of publicly available data on conflict minerals, we collected and analyzed stakeholder and government data covering 2003 through 2014. Because the data were not used to support findings, conclusions, or recommendations, we did not assess their reliability. To demonstrate the output of one of the in-region initiatives, we collected production and export data for the ITRI Supply Chain Initiative (iTSCi). We reviewed the production reports for three provinces in the DRC and for Rwanda published on iTSCi's website, abbreviated the data to include those most relevant to this study, and converted the data to tons and thousands of U.S. dollars. A limitation of the data is that the disaggregated production data–by mineral and mine–are proprietary to iSTCi members, so it is not possible to quantify the total amount of any of the minerals separately. We also collected conflict mineral production and export data from the websites of the governments of the DRC and Rwanda and from the International Trade Centre. We are presenting these data in the appendixes of the report because these are the only publicly available data we found for sourcing conflict minerals from the region. There is no distinction in these data between minerals that are conflict-free and those that have supported armed groups. Moreover, none of these data can be generalized or be used to infer the total production or export of conflict minerals from the DRC and the adjoining countries.

In response to a mandate in the Dodd-Frank Wall Street Reform and Consumer Protection Act that GAO submit an annual report that assesses the rate of sexual violence in war-torn areas of the DRC and adjoining countries, we identified and assessed any additional published information available on sexual violence in war-torn eastern DRC, as well as three adjoining countries that border the DRC—Rwanda, Uganda, and Burundi—since our 2013 report on sexual violence in these areas.[59] During the course of our review, we interviewed officials from State and USAID and interviewed NGO representatives and researchers to discuss the collection of sexual violence-related data—including population-based surveys and case file data—in the DRC and adjoining countries. Specifically, we followed up with researchers and representatives from those groups we interviewed for our prior review on sexual violence rates in eastern DRC and adjoining countries, including a representative from the Human Rights Center at the University of California, Berkeley, School of Law and others officials. The team also traveled to New York City to meet with officials from the United Nations Population Fund, United Nations High Commissioner for Refugees, and the United Nations Special Representative of the Secretary-General on Sexual Violence in Conflict. We also conducted Internet literature searches to identify new academic articles containing any additional information on sexual violence since our 2013 report.[60]

We conducted this performance audit from September 2013 to June 2014 in accordance with generally accepted government auditing standards. Those standards require that we plan and perform the audit to obtain sufficient, appropriate evidence to provide a reasonable basis for our findings and conclusions based on our audit objectives. We believe that the evidence obtained provides a reasonable basis for our findings and conclusions based on our audit objectives.

Appendix II. State Department Map of Armed Groups Present at Mine Sites in the Democratic Republic of the Congo

Figure 5 depicts the most recent map published by the Department of State. The map focuses on the exploitation of tin, tantalum, tungsten, and gold in the provinces of North and South Kivu and parts of Orientale, Maniema, and Katanga provinces.

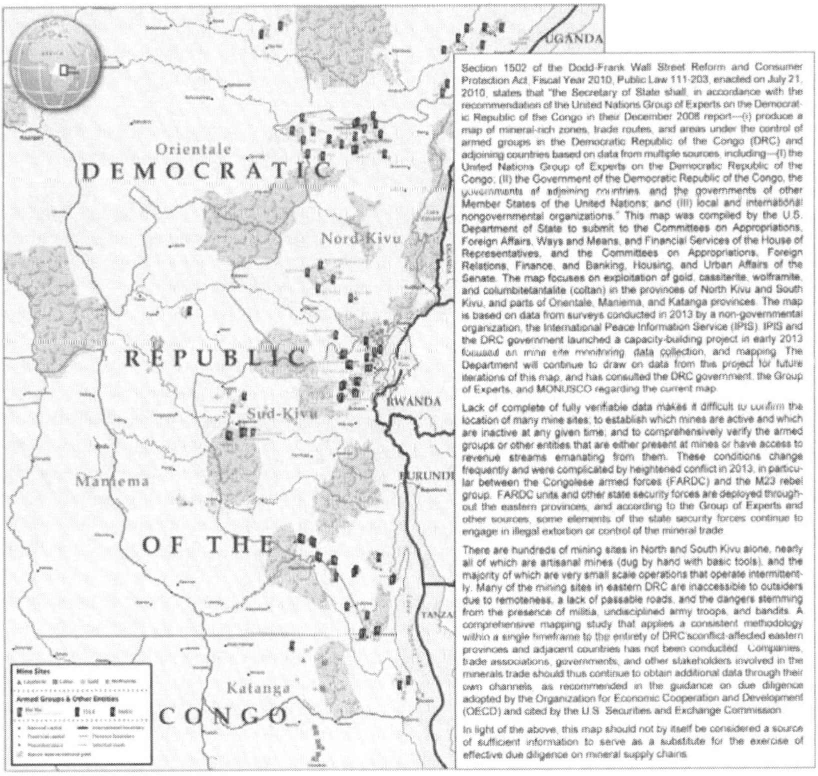

Source: State. | GAO-14-575

Figure 5. State Department Map of the Presence of Armed Groups at Mine Sites in the Democratic Republic of the Congo, as of February 2014.

Appendix III. SEC Flowchart Summary of the Disclosure Process for the Final Conflict Minerals Rule

When the SEC adopted the conflict minerals rule in August 2012, it published a flowchart summary of the final rule to guide SEC-reporting companies affected by the rule through the disclosure process (see figure 6). In general, the process reflects that an SEC-reporting company needs to (1) determine whether its manufactured products contain conflict minerals, (2) determine whether conflict minerals are necessary to the functionality or production of the product and if it originated in the DRC or an adjoining country, and (3) possibly conduct due diligence and potentially provide a Conflict Minerals Report.

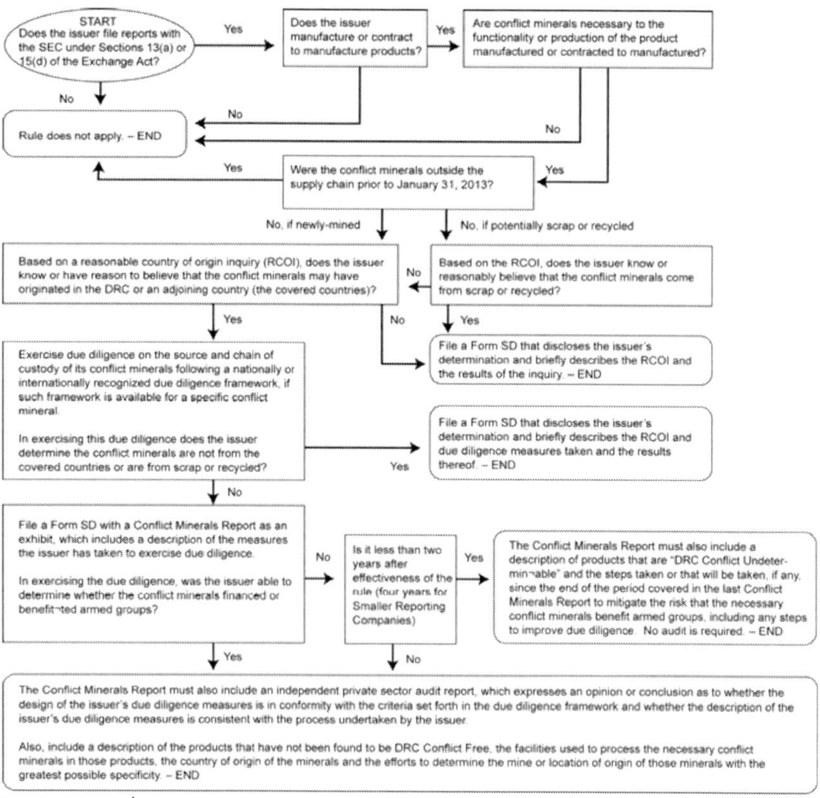

Source: SEC. | GAO-14-575

Note: SEC indicated that the flowchart is intended to be used as a guide and that issuers should refer to the text in the rule for a more comprehensive description of the rule's requirements. Further, SEC indicated that, as a result of the appeals court decision, SEC's stay order, and SEC staff's guidance, some of the steps outlined in the flowchart have been modified pending further action by SEC or a court.

Figure 6. SEC Flowchart Summary of Conflict Minerals Rule.

APPENDIX IV. SAMPLE DATA FROM THE ITRI SUPPLY CHAIN INITIATIVE

The ITRI Supply Chain Initiative (iTSCi) publishes qualitative and quantitative information on its projects in the DRC and Rwanda on its website. The information includes production data for minerals that have been mined and traded employing the iTSCi traceability system, audits of iTSCi member companies, and assessments of the political and security situation in various sites in the DRC and Rwanda. The mineral production and export data include minerals mined under the auspices of other stakeholder-led initiatives in the region. Almost all of the initiatives employ the iTSCi traceability system to track the mining and trading of minerals along the supply chain.[61]

One limitation of the data is that the production data–disaggregated by mineral and mine–are proprietary to iSTCi members. Therefore, it is not possible to quantify the total quantity or value of any of the minerals separately.[62]

Table 2. Production of Minerals from Maniema Covered by iTSCi for Q1 and Q2 of 2013, in Metric Tons

2013	Tin	Tin/ Tantalum[a]	Tin/ Tungsten[a]	Tantalum/ Tungsten[a]	Tungsten
Q1	17		11		1
Q2	225	1	17	1	
Total	**242**	**1**	**28**	**1**	**1**

Source: GAO based on iTSCi data. | GAO-14-575

Note: Empty cells indicate no recorded production.

[a] iTSCi data include traced material that are reported as combinations of more than one mineral.

Table 3. Sales of Minerals from Maniema Covered by iTSCi for Q1 and Q2 of 2013, in Thousands of U.S. Dollars

2013	Tin	Tin/ Tantalum[a]	Tin/ Tungsten[a]	Tantalum/ Tungsten[a]	Tungsten
Q1	$87		$46		$1
Q2	$1,163	$5	$84	$3	
Total	**$1,250**	**$5**	**$130**	**$3**	**$1**

Source: GAO based on iTSCi data. | GAO-14-575

Note: Empty cells indicate no recorded sales.

[a] iTSCi data include traced material that are reported as combinations of more than one mineral.

Tables 2-7 provide iTSCi production in metric tons and mineral sales in thousands of U.S. dollars for tin, tungsten, and tantalum coming from several provinces in the DRC, including Maniema, South Kivu, and Katanga.

Table 4. Production of Minerals from South Kivu Covered by iTSCi for Q4 of 2012 through Q2 of 2013, in Metric Tons

Year/ Quarter	Tin	Tin/ Tantalum[a]
2012/Q4	181	
2013/Q1	200	
2013/Q2	80	1
Total	**460**	**1**

Source: GAO based on iTSCi data. | GAO-14-575

Note: Empty cells indicate no recorded production. Totals may not add due to rounding.

[a] iTSCi data include traced material that are reported as combinations of more than one mineral.

Table 5. Sales of Minerals from South Kivu Covered by iTSCi for Q4 of 2012 through Q2 of 2013, in Thousands of U.S. Dollars

Year/ Quarter	Tin	Tin/ Tantalum[a]
2012/Q4	$760	
2013/Q1	$800	
2013/Q2	$286	$3
Total	**$1,847**	**$3**

Source: GAO based on iTSCi data. | GAO-14-575

Note: Empty cells indicate no recorded sales. Totals may not add due to rounding.

[a] iTSCi data include traced material that are reported as combinations of more than one mineral.

Table 6. Production of Minerals from Katanga Covered by iTSCi for Q2 of 2011 through Q2 of 2013, in Metric Tons

Year/ Quarter	Tin	Tin/ Tantalum[a]	Tin/ Tungsten[a]	Tantalum	Tantalum/ Tungsten[a]	Tungsten
2011/Q2	796	193	2	54		3
2011/Q3	1,436	234	2	51		16
2011/Q4	779	121	2	128		26
2012/Q1	753	252	15	114		12
2012/Q2	651	275	40	60	2	15
2012/Q3	671	258	18	64	2	40
2012/Q4	497	188	7	45		29
2013/Q1	340	211	7	34		9
2013/Q2	648	441	2	40	0.4	6
Total	**6,572**	**2,172**	**97**	**588**	**3**	**155**

Source: GAO based on iTSCi data. | GAO-14-575

Note: Empty cells indicate no recorded production. Totals may not add due to rounding.

[a] iTSCi data include traced material that are reported as combinations of more than one mineral.

Table 7. Sales of Minerals from Katanga Covered by iTSCi for Q2 of 2011 through Q2 of 2013, in Thousands of U.S. Dollars

Year/Quarter	Tin	Tin/ Tantalum[a]	Tin/ Tungsten[a]	Tantalum	Tantalum/ Tungsten[a]	Tungsten
2011/Q2	$5,166	$1,290	$14	$1,548		$21
2011/Q3	$9,839	$1,746	$18	$1,733		$122
2011/Q4	$4,072	$718	$40	$4,448		$152
2012/Q1	$4,224	$1,783	$67	$3,758		$70
2012/Q2	$3,701	$1,811	$193	$2,012	$7	$97
2012/Q3	$3,352	$1,720	$172	$2,114	$5	$181
2012/Q4	$2,906	$1,268	$106	$1,571		$105
2013/Q1	$2,112	$1,449	$38	$1,092		$74
2013/Q2	$3,744	$3,583	$11	$1,502	$2	$19
Total	**$39,117**	**$15,367**	**$659**	**$19,778**	**$15**	**$840**

Source: GAO based on iTSCi data. | GAO-14-575

Note: Empty cells indicate no recorded sales. Totals may not add due to rounding.

[a] iTSCi data include traced material that are reported as combinations of more than one mineral.

Tables 8 and 9 provide iTSCi production in tons and mineral sales in thousands of U.S. dollars for tin, tungsten and tantalum in Rwanda.

Table 8. Production of Minerals from Rwanda Covered by iTSCi for Q2 of 2011 through Q2 of 2013, in Metric Tons

Year/ Quarter	Tin	Tin/ Tantalum[a]	Tin/ Tungsten[a]	Tin/ Tantalum/ Tungsten[a]	Tantalum	Tantalum/ Tungsten[a]	Tungsten
2011/Q2	1,393	442	3	3	56	1	353
2011/Q3	1,318	406	1	4	56		438
2011/Q4	1,387	381	14	1	41		528
2012/Q1	825	316	8	3	102	3	575
2012/Q2	671	397	5	16	108	1	605
2012/Q3	780	509	4	26	162	1	582

Year/ Quarter	Tin	Tin/ Tantalum[a]	Tin/ Tungsten[a]	Tin/ Tantalum/ Tungsten[a]	Tantalum	Tantalum/ Tungsten[a]	Tungsten
2012/Q4	842	489	2	24	165	2	454
2013/Q1	915	678	1	3	178		538
2013/Q2	827	799	1		248		486
Total	**8,958**	**4,417**	**39**	**79**	**1,115**	**8**	**4,560**

Source: GAO based on iTSCi data. | GAO-14-575

Note: Empty cells indicate no recorded production. Totals may not add due to rounding.

[a] iTSCi data include traced material that are reported as combinations of more than one mineral.

Table 9. Sales of Minerals from Rwanda Covered by iTSCi for Q2 of 2011 through Q2 of 2013, in Thousands of U.S. Dollars

Year/ Quarter	Tin	Tin/ Tantalum[a]	Tin/ Tungsten[a]	Tin/ Tantalum/ Tungsten[a]	Tantalum	Tantalum/ Tungsten[a]	Tungsten
2011/Q2	$11,481	$6,217	$32	$25	$1,095	$8	$1,988
2011/Q3	$9,533	$4,778	$16	$26	$972		$2,935
2011/Q4	$9,333	$4,423	$154	$10	$720		$3,608
2012/Q1	$4,874	$4,383	$92	$26	$2,068	$17	$4,007
2012/Q2	$3,369	$4,980	$31	$98	$2,098	$15	$4,691
2012/Q3	$4,273	$6,394	$27	$176	$3,473	$12	$4,891
2012/Q4	$6,026	$8,008	$26	$263	$4,148	$35	$4,235
2013/Q1	$6,982	$11,031	$19	$59	$5,654		$6,678
2013/Q2	$5,988	$14,857	$24		$8,331		$4,605
Total	**$61,859**	**$65,070**	**$421**	**$683**	**$28,560**	**$87**	**$37,637**

Source: GAO based on iTSCi data. | GAO-14-575

Note: Empty cells indicate no recorded sales. Totals may not add due to rounding.

[a] iTSCi data include traced material that are reported as combinations of more than one mineral.

Appendix V. Conflict Mineral Production and Export Data as Reported by the DRC and Rwandan Governments

The DRC Ministry of Mines published data on the volume of production and exports for tin, tantalum, tungsten, and gold for 2003-2012 (see tables 10-13).[63]

Table 10. Total Volume of Tin Production and Export from DRC for 2003-2012, in Metric Tons

	Production	Export
2003	2,986	2,986
2004	2,317	2,945
2005	5,171	5,971
2006	5,528	2,388
2007	16,050	14,694
2008	20,013	19,189
2009	16,584	15,512
2010	11,943	13,415
2011	8,921	9,267
2012	7,189	8,018
Total	**96,702**	**94,385**

Source: GAO based on DRC Ministry of Mines. | GAO-14-575

Table 11. Total Volume of Tantalum Production and Export from DRC for 2003-2012, in Metric Tons

	Production	Export
2003	47	47
2004	78	32
2005	184	185
2006	31	24
2007	393	393
2008	530	531
2009	509	464
2010	492	440
2011	2,199	536
2012	574	586
Total	**5,037**	**3,238**

Source: GAO based on DRC Ministry of Mines. | GAO-14-575

Table 12. Total Volume of Tungsten Production and Export from DRC for 2003-2012, in Metric Tons

	Production	Export
2003	No data	100
2004	272	190
2005	405	311
2006	1,001	471
2007	1,265	1,194
2008	621	725
2009	458	365
2010	44	45
2011	84	87
2012	197	71
Total	**4,347**	**3,559**

Source: GAO based on DRC Ministry of Mines. | GAO-14-575

Table 13. Total Volume of Gold Production and Export from DRC for 2003-2012, in Metric Tons

	Production	Export
2003	No data	No data
2004	0.7	0.0
2005	0.6	0.6
2006	0.3	0.1
2007	0.1	0.1
2008	0.1	0.1
2009	0.2	0.2
2010	0.2	0.2
2011	0.3	0.2
2012	2.8	2.4
Total	**5.3**	**3.9**

Source: GAO based on DRC Ministry of Mines. | GAO-14-575

The Rwanda Natural Resources Authority published data on the volume and value of tin, tantalum and tungsten exported from Rwanda from January to August of 2013 (see tables 14 and 15).[64]

Table 14. Total Volume of Tin, Tantalum, and Tungsten Exports from Rwanda January-August 2013, in Metric Tons

	Tin	Tantalum	Tungsten
January	421	84	184
February	224	48	130
March	475	284	185
April	510	251	230
May	458	273	157
June	267	212	177
July	396	288	151
August	407	245	164
Total	**3,158**	**1,686**	**1,378**

Source: GAO based on Rwanda Natural Resources Authority. | GAO-14-575
Note: Totals may not add due to rounding.

Table 15. Total Value of Tin, Tantalum, and Tungsten Exports from Rwanda January-August 2013, in Thousands of U.S. Dollars

	Tin	Tantalum	Tungsten
January	$5,386	$4,953	$2,178
February	$2,803	$3,012	$1,564
March	$8,314	$16,379	$2,118
April	$6,352	$15,736	$3,915
May	$5,365	$16,583	$2,075
June	$3,206	$12,373	$2,559
July	$4,297	$17,822	$2,103
August	$4,509	$15,144	$2,277
Total	**$40,233**	**$102,002**	**$18,789**

Source: GAO based on Rwanda Natural Resources Authority. | GAO-14-575
Note: Totals may not add due to rounding.

APPENDIX VI. CONFLICT MINERALS DATA FROM THE INTERNATIONAL TRADE CENTRE DATA ON GLOBAL CONFLICT MINERAL IMPORTS

The International Trade Centre (ITC), a joint agency of the World Trade Organization and the United Nations (UN) compiles a Trade Map with data including the global exports in tons and export value of tin, tantalum, tungsten and gold from the DRC and adjoining countries, as available for fiscal years 2009-2013 (see tables 16-31 for ITC data as of June 2014).[65] The ITC calculated these data using the UN Commodity Trade Statistics Database, which compiles trade data from UN member countries.

These data do not provide a comprehensive depiction of the flow of conflict minerals exported from the DRC and adjoining countries; rather, they are an estimate based on imports data from reporting partner countries. Furthermore, these data may include imports of conflict minerals that have financed armed groups.[66]

In many instances, there are no data listed for a particular mineral or year. There were no data for import volume or value of conflict minerals in the ITC database for the Republic of the Congo (Congo-Brazzaville) or South Sudan.

Table 16. ITC Estimates of Exports of Conflict Minerals from the Democratic Republic of the Congo for Fiscal Years 2009-2013, in Metric Tons

	Tin	Tantalum[b]	Tungsten	Gold
FY2009	160[a]	121[a]	471[a]	No quantity[a]
FY2010	2,589[a]	252[a]	87[a]	0[a]
FY2011	510[a]	101[a]	10[a]	0[a]
FY2012	1,670[a]	326[a]		0[a]
FY2013	2,621[a]	329[a]	95[a]	0[a]
Total	**7,550**	**1,129**	**663**	0[a]

Source: GAO analysis of Trade Map data. | GAO-14-575

Note: According to the ITC, if there is no record in the export dataset for a specific mineral and a specific partner country, they consider that the country has not exported the mineral for that year and publish a zero value. "No quantity" means that the country under review has reported a value but no quantity for the product and partner under review. Empty cells indicate that no partner country has reported importation of that mineral from that country.

[a] According to the ITC, these estimates of exports have been reconstructed on the basis of data reported by importing partner countries, known as mirror statistics. The ITC states that mirror statistics are a "second-best solution" when no data exist but are subject to limitations, including the omission of non-reporting countries, particularly in Africa.

[b] Export data for tantalum ores and concentrates also include exports of niobium and vanadium ores and concentrates.

Table 17. ITC Estimates of Export Value of Conflict Minerals from the DRC for F Fiscal Years 2009-2013, in Thousands of U.S. Dollars

	Tin	Tantalum[b]	Tungsten	Gold
FY2009	$1,061[a]	$2,575[a]	$4,357[a]	$1,251[a]
FY2010	$23,045[a]	$5,161[a]	$806[a]	$1,140[a]
FY2011	$8,786[a]	$9,131[a]	$263[a]	$4,426[a]
FY2012	$17,719[a]	$24,006[a]		$1,394[a]
FY2013	$32,628[a]	$22,142[a]	$1,339[a]	$2,729[a]
Total	**$83,239**	**$63,015**	**$6,765**	**$10,940**

Source: GAO analysis of Trade Map data. | GAO-14-575

Note: According to the ITC, empty cells indicate that no partner country has reported importation of that mineral from that country.

[a] According to the ITC, these estimates of exports have been reconstructed on the basis of data reported by importing partner countries, known as mirror statistics. The ITC states that mirror statistics are a "second-best solution" when no data exist, but are subject to limitations, including the omission of non-reporting countries, particularly in Africa.

[b] Export data for tantalum ores and concentrates also include exports of niobium and vanadium ores and concentrates.

Table 18. ITC Estimates of Exports of Conflict Minerals from Angola for Fiscal Years 2009-2012, in Metric Tons

	Tin	Tantalum[b]	Tungsten	Gold
FY2009				
FY2010				
FY2011			No quantity[a]	0
FY2012		19[a]		
FY2013	No quantity[a]	28[a]		
Total		47		0

Source: GAO analysis of Trade Map data. | GAO-14-575

Note: According to the ITC, if there is no record in the export dataset for a specific mineral and a specific partner country, they consider that the country has not exported the mineral for that year and publish a zero value. "No quantity" means that the country under review has reported a value but no quantity for the product and partner under review. Empty cells indicate that no partner country has reported importation of that mineral from that country.

[a] According to the ITC, these estimates of exports have been reconstructed on the basis of data reported by importing partner countries, known as mirror statistics. The ITC states that mirror statistics are a "second-best solution" when no data exist

,but are subject to limitations, including the omission of non-reporting countries, particularly in Africa.

[b] Export data for tantalum ores and concentrates also include exports of niobium and vanadium ores and concentrates.

Table 19. ITC Estimates of Export Value of Conflict Minerals from Angola for Fiscal Years 2009-2013, in Thousands of U.S. Dollars

	Tin	Tantalum[b]	Tungsten	Gold
FY2009				
FY2010				
FY2011			$0[a]	$90[a]
FY2012		$31[a]		
FY2013	$0[a]	$7		
Total	**$0**	**$38**	**$0**	**$90**

Source: GAO analysis of Trade Map data. | GAO-14-575

Note: According to the ITC, if there is no record in the export dataset for a specific mineral and a specific partner country, they consider that the country has not exported the mineral for that year and publish a zero value. Empty cells indicate that no partner country has reported importation of that mineral from that country.

[a] According to the ITC, these estimates of exports have been reconstructed on the basis of data reported by importing partner countries, known as mirror statistics. The ITC states that mirror statistics are a "second-best solution" when no data exist, but are subject to limitations, including the omission of non-reporting countries, particularly in Africa.

[b] Export data for tantalum ores and concentrates also include exports of niobium and vanadium ores and concentrates.

Table 20. ITC Estimates of Exports of Conflict Minerals from Burundi for Fiscal Years 2009-2012, in Metric Tons

	Tin	Tantalum[a]	Tungsten	Gold
FY2009	26	0	264	1
FY2010	32	0	503	0
FY2011	0	17	330	1
FY2012	0	0	32	2
FY2013	No quantity	76	148	3
Total	**58**	**93**	**1,277**	**7**

Source: GAO analysis of Trade Map data. | GAO-14-575

Note: According to the ITC, if there is no record in the export dataset for a specific mineral and a specific partner country, they consider that the country has not exported the mineral for that year and publish a zero value. "No quantity" means that the country under review has reported a value but no quantity for the product and partner under review.

[a] Export data for tantalum ores and concentrates also include exports of niobium and vanadium ores and concentrates.

Table 21. ITC Estimates of Export Value of Conflict Minerals from Burundi for Fiscal Years 2009-2013, in Thousands of U.S. Dollars

	Tin	Tantalum[a]	Tungsten	Gold
FY2009	$175	$0	$1,547	$28,979
FY2010	$297	$0	$2,951	$13,003
FY2011	$15	$20	$4,607	$59,047
FY2012	$0	$0	$226	$105,196
FY2013	$0	$2,753	$1,776	$121,470
Total	**$487**	**$2,773**	**$11,107**	**$327,695**

Source: GAO analysis of Trade Map data. | GAO-14-575

Note: According to the ITC, if there is no record in the export dataset for a specific mineral and a specific partner country, they consider that the country has not exported the mineral for that year and publish a zero value.

[a] Export data for tantalum ores and concentrates also include exports of niobium and vanadium ores and concentrates.

Table 22. ITC Estimates of Exports of Conflict Minerals from Central African Republic for Fiscal Years 2009-2012, in Metric Tons

	Tin	Tantalum[a]	Tungsten	Gold
FY2009				24
FY2010				49
FY2011				21
FY2012				6
FY2013				691
Total				**791**

Source: GAO analysis of Trade Map data. | GAO-14-575

Note: According to the ITC, empty cells indicate that no partner country has reported importation of that mineral from that country.

[a] Export data for tantalum ores and concentrates also include exports of niobium and vanadium ores and concentrates.

Table 23. ITC Estimates of Export Value of Conflict Minerals from Central African Republic for Fiscal Years 2009-2013, in Thousands of U.S. Dollars

	Tin	Tantalum[a]	Tungsten	Gold
FY2009				$473
FY2010				$1,277
FY2011				$2,509
FY2012				$1,367
FY2013				$318
Total				**$5,944**

Source: GAO analysis of Trade Map data. | GAO-14-575

Note: According to the ITC, empty cells indicate that no partner country has reported importation of that mineral from that country.

[a] Export data for tantalum ores and concentrates also include exports of niobium and vanadium ores and concentrates.

Table 24. ITC Estimates of Exports of Conflict Minerals from Rwanda for Fiscal Years 2009-2012, in Metric Tons

	Tin	Tantalum[a]	Tungsten	Gold
FY2009	5,224	958	945	0
FY2010	6,477	827	879	0
FY2011	7,314	915	784	0
FY2012	4,658	1,139	1,372	0
FY2013	4,820	2,466	2,119	0
Total	**28,493**	**6,305**	**6,099**	**0**

Source: GAO analysis of Trade Map data. | GAO-14-575

Note: According to the ITC, if there is no record in the export dataset for a specific mineral and a specific partner country, they consider that the country has not exported the mineral for that year and publish a zero value.

[a] Export data for tantalum ores and concentrates also include exports of niobium and vanadium ores and concentrates.

Table 25. ITC Estimates of Export Value of Conflict Minerals from Rwanda for Fiscal Years 2009-2013, in Thousands of U.S. Dollars

	Tin	Tantalum[a]	Tungsten	Gold
FY2009	$34,235	$18,082	$6,283	$856
FY2010	$65,834	$20,034	$7,254	$170
FY2011	$101,921	$39,000	$11,677	0
FY2012	$52,912	$56,698	$21,354	0
FY2013	$57,957	$134,575	$28,866	0
Total	**$312,859**	**$268,389**	**$75,434**	**$1,026**

Source: GAO analysis of Trade Map data. | GAO-14-575

Note: According to the ITC, if there is no record in the export dataset for a specific mineral and a specific partner country, they consider that the country has not exported the mineral for that year and publish a zero value.

[a] Export data for tantalum ores and concentrates also include exports of niobium and vanadium ores and concentrates.

Table 26. ITC Estimates of Export of Conflict Minerals from Tanzania for Fiscal Years 2009-2012, in Metric Tons

	Tin	Tantalum[a]	Tungsten	Gold
FY2009	0	0	0	402
FY2010	1	0	0	325
FY2011	66	160	0	43
FY2012	71	112	0	319
FY2013	57	No quantity	No quantity	140
Total	**195**	**272**	**0**	**1,229**

Source: GAO analysis of Trade Map data. | GAO-14-575

Note: According to the ITC, if there is no record in the export dataset for a specific mineral and a specific partner country, they consider that the country has not exported the mineral for that year and publish a zero value. "No quantity" means that the country under review has reported a value but no quantity for the product and partner under review.

[a] Export data for tantalum ores and concentrates also include exports of niobium and vanadium ores and concentrates.

Table 27. ITC Estimates of Export Value of Conflict Minerals from Tanzania for Fiscal Years 2009-2013, in Thousands of U.S. Dollars

	Tin	Tantalum[a]	Tungsten	Gold
FY2009	$0	$0	$0	$818,598
FY2010	$5	$0	$0	$966,077
FY2011	$373	$59	$0	$1,718,196
FY2012	$351	$153	$0	$1,863,327
FY2013	$329	$0	$0	$1,549,580
Total	**$1,058**	**$212**	**$0**	**$6,915,778**

Source: GAO analysis of Trade Map data. | GAO-14-575

Note: According to the ITC, if there is no record in the export dataset for a specific mineral and a specific partner country, they consider that the country has not exported the mineral for that year and publish a zero value.

[a] Export data for tantalum ores and concentrates also include exports of niobium and vanadium ores and concentrates.

Table 28. ITC Estimates of Exports of Conflict Minerals from Uganda for Fiscal Years 2009-2012, in Metric Tons

	Tin	Tantalum[a]	Tungsten	Gold
FY2009	7	No quantity	0	1
FY2010	No quantity	No quantity	0	1
FY2011	0	No quantity	No quantity	0
FY2012	No quantity	No Quantity	100	0
FY2013	12	8	91	0
Total	19	8	191	2

Source: GAO analysis of Trade Map data. | GAO-14-575

Note: According to the ITC, if there is no record in the export dataset for a specific mineral and a specific partner country, they consider that the country has not exported the mineral for that year and publish a zero value. "No quantity" means that the country under review has reported a value but no quantity for the product and partner under review.

[a] Export data for tantalum ores and concentrates also include exports of niobium and vanadium ores and concentrates.

Table 29. ITC Estimates of Export Value of Conflict Minerals from Uganda for Fiscal Years 2009-2013, in Thousands of U.S. Dollars

	Tin	Tantalum[a]	Tungsten	Gold
FY2009	$2	$0	$0	$13,100
FY2010	$0	$0	$0	$30,072
FY2011	$1	$0	$0	$6,795
FY2012	$0	$0	$979	$9,163
FY2013	$128	$22	$972	$3,128
Total	**$131**	**$22**	**$1,951**	**$62,258**

Source: GAO analysis of Trade Map data. | GAO-14-575

Note: According to the ITC, if there is no record in the export dataset for a specific mineral and a specific partner country, they consider that the country has not exported the mineral for that year and publish a zero value.

[a] Export data for tantalum ores and concentrates also include exports of niobium and vanadium ores and concentrates.

Table 30. ITC Estimates of Exports of Conflict Minerals from Zambia for Fiscal Years 2009-2012, in Metric Tons

	Tin	Tantalum[a]	Tungsten	Gold
FY2009	27	No quantity	0	695
FY2010	0	No quantity	23	555
FY2011	28	0	72	2
FY2012	0	No quantity	0	1,492
FY2013	6	0	0	4
Total	**61**	**0**	**95**	**2,748**

Source: GAO analysis of Trade Map data. | GAO-14-575

Note: According to the ITC, if there is no record in the export dataset for a specific mineral and a specific partner country, they consider that the country has not exported the mineral for that year and publish a zero value. "No quantity" means that the country under review has reported a value but no quantity for the product and partner under review.

[a] Export data for tantalum ores and concentrates also include exports of niobium and vanadium ores and concentrates.

Table 31. ITC Estimates of Export Value of Conflict Minerals from Zambia for Fiscal Years 2009-2013, in Thousands of U.S. Dollars

	Tin	Tantalum[a]	Tungsten	Gold
FY2009	$77	$0	$0	$19,004
FY2010	$1	$0	$12	$46,411
FY2011	$50	$0	$6	$81,040
FY2012	$0	$0	$0	$140,744
FY2013	$4	$0	$0	$164,079
Total	**$132**	**$0**	**$18**	**$451,278**

Source: GAO analysis of Trade Map data. | GAO-14-575

Note: According to the ITC, if there is no record in the export dataset for a specific mineral and a specific partner country, they consider that the country has not exported the mineral for that year and publish a zero value.

[a] Export data for tantalum ores and concentrates also include exports of niobium and vanadium ores and concentrates.

End Notes

[1] GAO, *The Democratic Republic of the Congo: U.S. Agencies Should Take Further Actions to Contribute to the Effective Regulation and Control of the Minerals Trade in Eastern Democratic Republic of the Congo*, GAO-10-1030 (Washington, D.C.: Sept. 30, 2010).

[2] M-23 was created in May 2012 in eastern DRC and surrendered to DRC forces in November 2013.

[3] United Nations Security Council, "Letter dated 12 December 2013 from the Group of Experts on the Democratic Republic of the Congo addressed to the Chair of the Security Council Committee established pursuant to resolution 1533 (2004) concerning the Democratic Republic of the Congo," Annex to "Letter dated 22 January 2014 from the Coordinator of the Group of Experts on the Democratic Republic of the Congo addressed to the President of the Security Council," S/2014/42 (Jan. 23, 2014).

[4] The Dodd-Frank Act defines conflict minerals as columbite-tantalite (coltan), cassiterite, gold, wolframite, or their derivatives, or any other mineral or its derivatives that are determined by the Secretary of State to be financing conflict in the DRC or an adjoining country. See Pub. L. No. 111-203 §1502, (e)(4). Columbite-tantalite (coltan), cassiterite, and wolframite are the ores from which tantalum, tin, and tungsten, respectively, are processed.

[5] In this report, the countries adjoining DRC—Angola, Burundi, Central African Republic, the Republic of the Congo, Rwanda, South Sudan, Tanzania, Uganda, and Zambia—are often termed "Dodd-Frank affected countries."

[6] Under the rule, SEC-reporting companies with "necessary" conflict minerals in products they manufactured or contracted to have manufactured must conduct a "reasonable country of origin inquiry" to determine whether their conflict minerals originated in the DRC or adjoining countries. If none of their conflict minerals originated in that region, they must file a Form SD to disclose the reasonable country of origin inquiry they conducted and the results. If any of a company's "necessary" conflict minerals originated in the DRC or

adjoining countries, the company must exercise due diligence on the source and chain of custody of its conflict minerals and also file a Conflict Minerals Report.

[7] GAO, *The Democratic Republic of the Congo: Information on the Rate of Sexual Violence in War-Torn Eastern DRC and Adjoining Countries*, GAO-11-702 (Washington, D.C.: July 13, 2011); *Conflict Minerals Disclosure Rule: SEC's Actions and Stakeholder-Developed Initiatives*, GAO-12-763 (Washington, D.C.: July 16, 2012); *SEC Conflict Minerals Rule: Information on Responsible Sourcing and Companies Affected*, GAO-13-689 (Washington, D.C.: July 18, 2013). In our 2012 report, we addressed, for example, a congressional mandate that asked us to describe any issues that the SEC encountered in promulgating a conflict minerals disclosure rule. In our 2013 report, we addressed, for example, a mandate that asked us to describe information that may be publicly available about entities that use conflict minerals but are not required to report to SEC under the rule.

[8] Stakeholders include relevant U.S. agencies; foreign governments; multilateral organizations, such as the Organization for Economic Cooperation and Development (OECD); nongovernmental organizations (NGOs); and industry associations.

[9] GAO-13-689.

[10] GAO-13-689.

[11] GAO-13-689.

[12] The Office of the High Commissioner for Human Rights and the United Nations Organization Stabilization Mission in the Democratic Republic of the Congo (MONUSCO) issued a report in May 2013 on mass rapes and other violations of human rights committed in Minova, a town in South Kivu, and surrounding areas in November 2012. The report documented 135 cases of sexual violence perpetrated by elements of the Congolese national army in and around the town of Minova as units retreated from the front lines.

[13] The eleven signatories to the framework were Angola, Burundi, the Central Africa Republic, the DRC, the Republic of the Congo (also known Congo-Brazzaville), Rwanda, South Africa, South Sudan, Tanzania, Uganda, and Zambia. The UN Secretary-General, the Chairperson of the African Union Commission, the Chairperson of the Southern African Development Community, and the Chairperson of the International Conference on the Great Lakes Region signed the agreement as witnesses. In January 2014, two additional countries, Kenya and Sudan, also signed the framework. According to the UN, the framework encompasses commitments at the national, regional and international levels to bring peace and stability to the eastern DRC and the region.

[14] According to the UN, internally displaced persons are people who have not crossed an international border but have moved to a different region than the one they call home within their own country to escape war, persecution, or terror.

[15] Pub. L. No. 109-456.

[16] Initially, MONUC's focus was on the ceasefire and disengagement of forces and maintenance of liaison with all parties involved with the civil war, but the operation expanded to include the effective protection of civilians, humanitarian personnel, and human rights defenders under imminent threat of physical violence.

[17] According to the UN, the envoy's key tasks include undertaking good offices to strengthen the relations between the signatories of the framework, revitalizing existing accords and coordinating the international engagement.

[18] The mandate of the force intervention brigade was renewed by the UN Security Council until March 2015.

[19] Canada also has a proposed conflict minerals initiative. According to State, Canada's Conflict Minerals Act was reintroduced for discussion in the Canadian parliament in April 2014 and would require Canadian companies to exercise due diligence in respect of the exploitation and trading of designated conflict minerals originating in the Great Lakes Region of Africa.

[20] The OECD due diligence guidance, which OECD adopted in 2011, includes supplements on tin, tantalum, tungsten, and gold. The framework's five steps are (1) establishing strong company management systems, (2) identifying and assessing risk in the supply chain, (3)

designing and implementing a strategy to respond to identified risks, (4) carrying out an independent third-party audit of supply chain due diligence at identified points in the supply chain, and (5) reporting on supply chain due diligence. Organization for Economic Cooperation and Development, *OECD Due Diligence Guidance for Responsible Supply Chains of Minerals from Conflict-Affected and High-Risk Areas: Second Edition*, OECD Publishing (2013), accessed June 23, 2014, http://dx.doi.org/10.1787/9789264185050-en.

[21] Artisanal mining is a form of mining characterized by a lack of mechanization or capital investment.

[22] According to a World Gold Council representative, in most cases refiners are paid a fee to refine gold and the transaction is conducted by the miner, the trader, and the bank.

[23] See GAO-12-763.

[24] Organization for Economic Cooperation and Development, *OECD Due Diligence Guidance for Responsible Supply Chains of Minerals from Conflict-Affected and High-Risk Areas: Second Edition* (Paris: November 2012).

[25] The Project Management Institute, *The Standard for Program Management* (2013).

[26] The Act states that a strategy be submitted to Congress no later than 180 days after the enactment of the law.

[27] The Act requires that State should update its map every 180 days.

[28] IPIS has also published a report that analyzes the information it collects and populates in its interactive map. See International Peace Information Service, *Analysis of the Interactive Map of Artisanal Mining Areas in Eastern DR Congo,* (Antwerp, Belgium: November 2013), accessed June 23, 2014, http://ipisresearch.be/mapping/webmapping/drcongo.

[29] GAO-13-689.

[30] In October 2012, the U.S. Chamber of Commerce, the National Association of Manufacturers, and the Business Roundtable filed a lawsuit against SEC challenging the final conflict minerals rule, making claims based on the Administrative Procedure Act, the Securities Exchange Act of 1934, and the First Amendment. In July 2013, the U.S. District Court for the District of Columbia denied the plaintiff's claims on all counts. See *National Association of Manufacturers v. SEC*, 956 F. Supp. 2d 43 (D.D.C. 2013). In August 2013, the petitioners appealed the decision to the U.S. Court of Appeals for the District of Columbia Circuit, and on April 14, 2014, the appeals court upheld the District Court's findings on all the petitioners' claims except the First Amendment claims, concluding that section 13(p) of the Exchange Act and the final conflicts minerals rule violate the First Amendment "to the extent the statute and rule require regulated entities to report to the Commission and to state on their website that any of their products have 'not been found to be DRC conflict free.'" *National Association of Manufacturers, et al. v. SEC et al., No. 13-5252 (D.C. Cir. April 14, 2014).* In response to the appeals court's findings, SEC staff, on April 29, 2014, issued a statement that it expects companies to file any reports required under Rule 13p-1 subject to any further action that may be taken either by the Commission or a court. The SEC staff's statement contains guidance to companies, which provides, among other things, that no company is required to describe its products (1) as "DRC conflict free" (2) as having "not been found to be 'DRC conflict free'," or (3) as "DRC conflict undeterminable" in their reports. The guidance also states that a company may voluntarily elect to describe any of its products as "DRC conflict free" in its report if it had obtained an independent private sector audit as required by the rule. In addition, the guidance states that, pending further action, an independent private sector audit will not be required unless a company voluntarily elects to describe a product as "DRC conflict free" in its Conflict Minerals Report. On May 2, 2014, the SEC issued an order staying the effective date for compliance with the portions of Rule 13p-1 and Form SD subject to the appeals court's First Amendment holding pending the completion of judicial review. On May 5, 2014, the plaintiffs filed a motion with the appeals court asking the court to stay the entire rule pending the completion of judicial review which the Commission opposed, and on May 2014, the appeals court denied the motion.

[31] Pub. L. No. 111-203, sec 1502(b). The Act required that a rule should be promulgated no later than 270 days after the law's enactment. As we reported in 2012, SEC's issuance of the rule encountered delays, including delays in developing, modifying, and finalizing the rule, as a result of several factors. See GAO-12-763. Also see appendix III for a flowchart summarizing the conflict minerals rule that SEC published when the rule was adopted in 2012.

[32] The SEC rule indicates that if the deadline for filing the conflict mineral disclosure report occurs on a weekend, or on a holiday which the SEC is not open for business, then the deadline shall be the next business day.

[33] According to the rule, for a temporary period following the effective date of the rule—4 years for smaller reporting companies or 2 years for all other reporting companies—if a company is unable to determine whether the minerals in its products originated in DRC or the adjoining countries or financed or benefited armed groups in those countries, then those products are considered "DRC conflict undeterminable" and no audit is required. According to SEC officials, as a result of the SEC's partial stay order and the SEC staff's statement, companies are not required to describe their products in their reports as "DRC conflict undeterminable" pending further action by the SEC or a court, and no audits are required unless a company voluntarily describes any of its products as "DRC conflict free" in its Conflict Minerals Report.

[34] The Act also requires Commerce to annually report (1) an assessment of the accuracy of the independent private sector audits and other due diligence processes required under section 1502 and (2) recommendations for the processes to carry out such audits, including ways to improve the accuracy of the audits and to establish standards of best practices. Commerce officials stated that it was unable to implement these aspects of the Act's requirement because the SEC disclosure reports, which are needed for the assessment, were not due to SEC until June 2, 2014.

[35] Our 2013 report identified 82 smelters and refiners of tin, tantalum, tungsten, and gold located in China, of the 278 smelters and refiners that we identified worldwide.

[36] According to USAID, it holds two of the four government positions on the Committee in addition to State and the ICGLR. There are four civil society and four private sector members of the Committee as well.

[37] According to an NGO report, the objective of mineral trading centers, or *centre de negoce*, is to provide a center where miners and traders can do business without interference from armed groups. The report notes that, at the same time, state agents can exert control and levy taxes, traders can receive the necessary paperwork, and miners can profit from a market environment in which they can negotiate better prices for their products.

[38] The Regional Steering Committee, comprising technical experts from all ICGLR member states, has been charged with the steering of all activities within the initiative since September 2009. Its suggestions on policies and technical guidelines are to be approved at a later stage by the Regional Inter-Ministerial Committee as well as the ICGLR Summit.

[39] GAO-13-689.

[40] GAO-13-689.

[41] CFSI was founded in 2008 by members of the Electronic Industry Citizenship Coalition (EICC) and the Global e-Sustainability Initiative (GeSI). According to EICC and GeSI, as of April 30, 2013, CFSI now includes EICC and GeSI activities and is an expanded initiative of the EICC and GeSI Extractives Working Group, which includes more stakeholders and a wider range of industry sectors supporting the sourcing of conflict-free minerals.

[42] The actual number of smelters and refiners of conflict minerals is unknown. However, in 2013, we reported that the total number of smelters and refiners worldwide could be as high as 500. See GAO-13-689.

[43] The World Bank defines "civil society" as referring to the wide array of nongovernmental and not-for-profit organizations that have a presence in public life, expressing the interests and

values of their members or others, based on ethical, cultural, political, scientific, religious, or philanthropic considerations.

[44] LBMA oversees the London Good Delivery List, a list of 67 accredited gold refiners that meet a specific standard for the quality of their refined gold and silver bars. According to LBMA representatives, most banks and exchanges will only contract with refiners on the Good Delivery List.

[45] The Responsible Jewellery Council launched its chain-of-custody program to also help its member companies identify and track the supply chains of other precious metals, such as platinum, palladium, and rhodium. Chain of custody refers to the paper trail that documents the sequence of entities with custody of minerals as they move through a supply chain.

[46] Officials of the Responsible Jewellery Council reported that the council has developed cross-recognition agreements with CFSI, LBMA, and the World Gold Council to reduce unnecessary duplicative audits for their member companies. For example, according to an LBMA official, once LBMA approves a third-party audit, verifying that a gold refiner is conflict-free, the association contacts CFSI to add the refiner to CFSI's list of compliant smelters and refiners.

[47] According to iTSCi, as of November 2013, the traceability program is used at more than 850 mines in the South Kivu, Maniema, and Katanga provinces of the DRC and in Rwanda. Almost all in-region responsible sourcing initiatives—including Solutions for Hope and the Conflict-Free Tin Initiative—use iTSCi's traceability system of tracking minerals from the mine to the exporter to ensure that their minerals are conflict-free.

[48] According to Solutions for Hope, its closed pipeline involved working through a defined set of key suppliers—mines (including artisanal cooperatives), smelter/processor, component manufacturer, and end user—identified in advance of initiating the project.

[49] According to Solutions for Hope officials, a tin smelter in the Katanga province is currently idle due to lack of consistent power, and another smelter in Rwanda, which is working toward certification by the Conflict-Free Smelter Program, has been idle for over 5 years.

[50] According to an iTSCi official, the governance assessments also include summaries of all incidents that have occurred and how they have been mitigated or otherwise resolved.

[51] ICF International implements the Demographic and Health Surveys (DHS) Program, which has provided technical assistance to more than 300 surveys in over 90 countries. The DHS Program provides capacity building to host-country implementing agencies through all survey stages, including survey design and sampling, training, field work, data tabulation and analysis, report writing, and dissemination and use of findings.

[52] U.S. Department of State, Bureau of Democracy, Human Rights and Labor, *Country Reports on Human Rights Practices for 2013*, accessed May 18, 2014, http://www.state.gov/j/drl/rls/hrrpt/humanrightsreport/index.htm#wrapper.

[53] UN Security Council, *Report of the Secretary-General on the United Nations Organization Stabilization Mission in the Democratic Republic of the Congo*, S/2013/757 (New York, NY: Dec 17, 2013).

[54] *Progress and Obstacles in the Fight Against Impunity for Sexual Violence in the Democratic Republic of the Congo*, United Nations Joint Human Rights Office, April 2014.

[55] *Extent of Sexual Violence in the DRC and Actions to Fight against the Phenomenon of 2011 to 2012,* DRC Ministry of Gender, the Family, and Child; Kinshasa, June 2013. This report was not publicly available before we issued our report in July 2013.

[56] Medecins sans Frontieres, *Everyday Emergency: Silent Suffering in the Democratic Republic of Congo*, Amsterdam, 2014.

[57] GAO, *SEC Conflict Minerals Rule: Information on Responsible Sourcing and Companies Affected*, GAO-13-689 (Washington, D.C.: July 18, 2013).

[58] GAO, *Conflict Minerals Disclosure Rule: SEC's Actions and Stakeholder-Developed Initiatives*, GAO-12-763 (Washington, D.C.: July 16, 2012) and GAO-13-689.

[59] GAO-13-689. We did not include in our current report information from preliminary studies or reports.

[60] GAO-13-689.

[61] Because the data were not used to support findings, conclusions, or recommendations, we did not assess their reliability.

[62] Detailed data showing production and exports of each of the conflict minerals are proprietary to iTSCi members.

[63] The DRC Ministry of Mines statistics also include production and export data for other minerals. Because the data were not used to support findings, conclusions, or recommendations, we did not assess their reliability.

[64] Because the data were not used to support findings, conclusions, or recommendations, we did not assess their reliability.

[65] The Trade Map (or Market Access Map, Investment Map and Standards Map, respectively), International Trade Centre, www.intracen.org/marketanalysis.

[66] Because the data were not used to support findings, conclusions, or recommendations, we did not assess their reliability.

In: Conflict Minerals
Editor: Gary L. Simmons
ISBN: 978-1-63463-326-0

Chapter 2

SEC Conflict Minerals Rule: Information on Responsible Sourcing and Companies Affected*

United States Government Accountability Office

Why GAO Did This Study

The eastern part of the DRC has experienced recurring conflicts involving armed groups that have resulted in severe human rights abuses. In addition, armed groups have profited from the exploitation of minerals. In 2010, Congress enacted Section 1502(b) of the Dodd-Frank Wall Street Reform and Consumer Protection Act to address the exploitation of conflict minerals, which include tin, tantalum, tungsten, and gold, and the extreme levels of violence in the DRC. As required by Section 1502(b), the SEC issued a rule in August 2012 that requires companies to disclose their use of conflict minerals and the origin of those minerals. The act requires GAO to report on the rule's effectiveness, among other issues, beginning in 2012 and annually thereafter.

Initial company disclosure reports to SEC that would enable GAO to assess the effectiveness of the rule will not be due until May 2014. This report describes, among other issues, (1) factors that may impact whether SEC's rule denies armed groups in the DRC benefits from conflict

* This is an edited, reformatted and augmented version of a United States Government Accountability Office publication, No. GAO-13-689, dated July 2013.

minerals and (2) information about companies that use conflict minerals and are not required to report to SEC under the rule.

GAO reviewed and analyzed documents and interviewed representatives from SEC, the Department of State, the U.S. Agency for International Development, industry associations, NGOs, consulting firms, and international organizations. GAO also analyzed smelter and refiner information. This report does not contain recommendations.

WHAT GAO FOUND

Stakeholder-developed initiatives may facilitate companies' compliance with the Securities and Exchange Commission's (SEC) final conflict minerals rule, but other factors may affect the rule's impact on reducing benefits to armed groups in the Democratic Republic of the Congo (DRC) and neighboring countries. Agency and industry officials as well as representatives from international organizations and nongovernmental organizations (NGO) stated that adoption of the rule as well as stakeholder-developed initiatives—which include the development of guidance documents, audit protocols, and in-region sourcing of conflict minerals—can support companies' efforts to conduct due diligence and to identify and responsibly source conflict minerals. For example, officials GAO interviewed explained that the Conflict-Free Smelter Program enables suppliers to source conflict minerals from smelters (companies that refine the ore of the conflict minerals into metals) that have been certified by an independent third-party auditor as obtaining their minerals from sources that did not benefit armed groups. However, officials GAO interviewed cited constraining factors such as lack of security, lack of infrastructure, and lack of capacity in the DRC that could affect the ability to expand on efforts to achieve conflict-free sourcing of minerals from eastern DRC and thereby potentially contribute to armed groups' benefiting from the conflict minerals trade. For example, officials GAO interviewed noted that there is a lack of infrastructure in place that would enable companies to set up or expand operations in the DRC. Limited transportation and poor roads in eastern DRC also make it difficult to get to mine sites. Moreover, according to officials, the remoteness of mines also makes it difficult for DRC officials to validate mines and ensure that the mines have not been compromised by illegal armed groups.

Companies that are not required to file disclosures under SEC's conflict minerals rule may be affected by the rule. These companies may supply components or parts that contain conflict minerals to companies that report to

SEC under the rule, many of which could be original equipment manufacturers and component parts manufacturers. Estimates provided by public commentators responding to the rule indicate that roughly 280,000 suppliers could provide products to roughly 6,000 companies that report to the SEC under the rule and may be asked to provide information on their use of conflict minerals and the origin of the minerals as part of the rule's due diligence requirements. GAO found little available aggregated information about companies that do not report to SEC under the rule. However, GAO found that for smelters and refiners there is some aggregated information, such as the types of conflict minerals they use and their location. For example, GAO found that over half of the 278 smelters and refiners of conflict minerals it identified were located in Asia, many processed tin, and most did not have a conflict minerals policy publicly available.

ABBREVIATIONS

DHS	Demographic and Health Survey
Dodd-Frank Act	Dodd-Frank Wall Street Reform and Consumer Protection Act
DRC	Democratic Republic of the Congo
EICC	Electronic Industry Citizenship Coalition
GeSI	Global e-Sustainability Initiative
ICGLR	International Conference of the Great Lakes Region
iTSCi	ITRI Tin Supply Chain Initiative
LBMA	London Bullion Market Association
M23	March 23 Movement
MONUSCO	United Nations Organization Stabilization Mission in the Democratic Republic of the Congo
NGO	nongovernmental organization
OECD	Organisation for Economic Co-operation and Development
PPA	Public-Private Alliance for Responsible Minerals Trade
SEC	Securities and Exchange Commission
State	United States Department of State
UN	United Nations
UNGoE	United Nations Group of Experts
USAID	United States Agency for International Development

July 18, 2013

Congressional Committees

The Democratic Republic of the Congo (DRC) has long been the site of one of the world's worst humanitarian crises. Since 1998, over 5.4 million people have died as a result of the conflict which has also destabilized the eastern part of the country, created insecurity, displaced people, and perpetuated the cycle of poverty, according to estimates by the International Rescue Committee. In 2010 we reported that illegal armed groups, as well as some units of the Congolese national military, had continued to commit severe human rights abuses, including mass killings.[1] In 2013 the United Nations (UN) reported continuing violence by illegal armed groups, including violence by the group known as M23.[2] Sexual violence has also been a feature of the conflict in the DRC at least since the Congolese civil war of the mid-1990s. The illegal armed groups and units of the Congolese national military committing these atrocities also profit from the illegal mining of minerals. In response to the humanitarian situation in the DRC, Congress included in the Dodd-Frank Wall Street Reform and Consumer Protection Act (hereafter referred to as the Dodd-Frank Act, or the Act) provisions pertaining to the trade of DRC conflict minerals—tantalum, tin, tungsten, and gold.[3] Specifically, section 1502(a) of the Act states that "It is the sense of the Congress that the exploitation and trade of conflict minerals originating in the [DRC] is helping to finance conflict characterized by extreme levels of violence in the eastern [DRC], particularly sexual- and gender-based violence, and contributing to an emergency humanitarian situation therein." Section 1502(b) of the Act required the Securities and Exchange Commission (SEC) to issue a conflict minerals disclosure rule that requires issuers with conflict minerals that are necessary to the functionality or production of a product manufactured by such persons to file reports annually with SEC to disclose whether any of those minerals originated in the DRC or an adjoining country, and if so, to provide an additional report.[4] In response to section 1502 of the Act, SEC adopted the final conflict minerals rule on August 22, 2012.[5] The Act also required us to report, beginning in 2012 and annually thereafter, on the effectiveness of the rule in promoting peace and security in the DRC and adjoining countries; to describe information that may be publicly available about entities that use conflict minerals but are not required to report to SEC under the rule; and report, annually beginning in 2011, on the rate of sexual violence in war-torn areas of the DRC and neighboring countries. To address the mandate, in July 2011 we issued our first report on sexual violence[6] and in July 2012 we issued our second report, which focused on SEC's actions and stakeholder-developed initiatives involving conflict minerals and updated information on the rate of sexual violence.[7]

As in our 2012 report, we did not address the effectiveness of SEC's conflict minerals disclosure rule, as required under the legislation, because the first disclosures of companies' use of conflict minerals will not be due to SEC until May 2014 and sufficient time must elapse to allow the full impact of the rule to materialize. However, the effectiveness of the rule in promoting peace and security in the DRC will depend in part on its ability to limit funding by restricting the use of conflict minerals to illegal armed groups in the region to which those groups have access. As a result, this report describes (1) factors that may impact whether SEC's conflict minerals rule denies armed groups in the DRC and adjoining countries benefits from conflict minerals; (2) available information about entities that use conflict minerals and are not required to report to SEC under the rule;[8] and (3) any additional information available on the rate of sexual violence in eastern DRC and neighboring countries since our 2012 report.[9]

To address these objectives, we reviewed and analyzed reports and other documents from relevant U.S. agencies; multilateral organizations, such as the Organisation for Economic Co-operation and Development (OECD); nongovernmental organizations (NGO); and industry associations. We also analyzed information on tin, tantalum, tungsten, and gold smelters and refiners to ascertain the location of these entities as well as whether they had posted any conflict-free sourcing policies on their companies' websites. We interviewed officials, who are cognizant of conflict minerals issues, from SEC, the Department of State (State), and the United States Agency for International Development (USAID), as well as representatives from international organizations, NGOs, industry associations, consulting firms, and smelters and refiners of tin, tantalum, tungsten, and gold.[10] We chose the experts and stakeholders we interviewed to capture a range of perspectives about the types of minerals traded and because we had established contacts with these entities on our last review. These experts and stakeholders constitute a nongeneralizable sample. The information gathered cannot be generalized and cannot be used to infer views of other experts or stakeholders cognizant of conflict minerals issues. In the interviews, we asked the experts and stakeholders to provide factors that they believe may impact whether SEC's conflict minerals rule denies armed groups in the DRC benefits from conflict minerals. We identified 278 smelters and refiners of tin, tantalum, tungsten, and gold, and analyzed publicly available information on their practices and policies for sourcing conflict minerals, including any due diligence guidance the smelters or refiners reportedly followed for sourcing conflict minerals. We identified 278 smelters and refiners in our analysis; the total number of smelters and refiners is believed to be nearly 500, worldwide. The number of gold refiners could potentially be larger, considering that little equipment and space is required to refine gold, and it can be refined at the mine site. The 278 smelters and refiners we were able to identify may not be representative of others, and the information we report about these 278 cannot be generalized to other smelters and refiners of tin, tantalum, tungsten, and gold. See appendix I for a complete description of our scope and methodology.

We conducted this performance audit from November 2012 to July 2013 in accordance with generally accepted government auditing standards. Those standards require that we plan and perform the audit to obtain sufficient, appropriate evidence to provide a reasonable basis for our findings and conclusions based on our audit objectives. We believe that the evidence obtained provides a reasonable basis for our findings and conclusions based on our audit objectives.

BACKGROUND

History of the DRC: Conflict and Instability

The mineral-rich DRC, Africa's second-largest country, has been plagued by cycles of violence and instability. Since 1998, violent conflicts, poverty, and disease have killed more than 5.4 million people in the country, according to estimates by the International Rescue Committee. The DRC was colonized as a personal possession of Belgian King Leopold II in 1885 and administered by the Belgian government starting in 1907. It achieved independence from Belgium in 1960. For almost 30 years of the post-independence period, the DRC, then known as Zaire, was ruled by an authoritarian regime under Mobutu Sese Seko. Following the 1994 genocide in Rwanda and the establishment of a new government there, some perpetrators of the genocide and refugees fled to the neighboring Kivu provinces of eastern DRC. A rebellion began there in 1996, pitting the forces led by Laurent Kabila against the army of President Mobutu Sese Seko. Kabila's forces, aided by Rwanda and Uganda, took the capital city of Kinshasa in 1997 and renamed the country the Democratic Republic of the Congo. See figure 1 for a map of the DRC's provinces and neighboring countries.

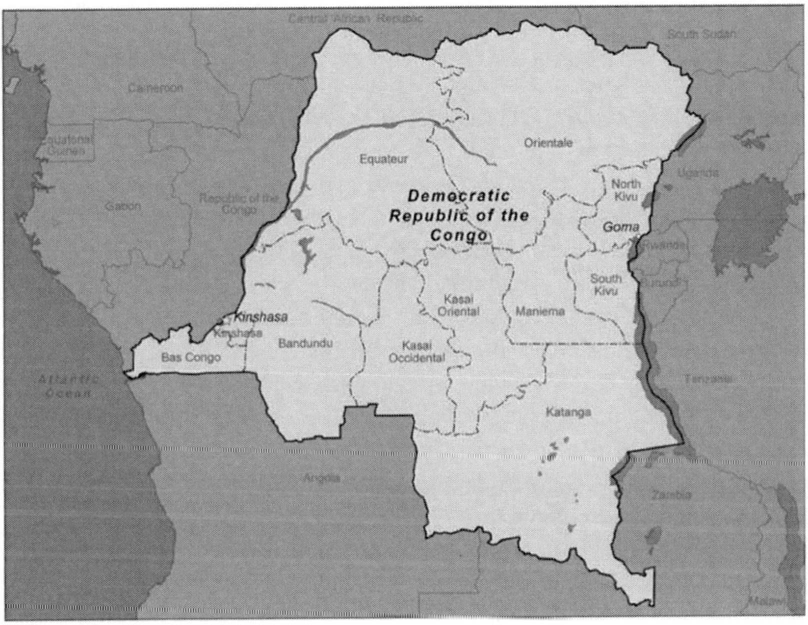

Source: GAO; United Nations (map).

Figure 1. Map of DRC with Provinces and Neighboring Countries.

A period of civil war among rival rebel groups ensued. In 2001 Laurent Kabila was assassinated and leadership shifted to his son Joseph Kabila, while the civil war continued. Starting in 1999 the UN Security Council authorized peacekeeping operations in the DRC which have been operating as the United Nations Organization Stabilization Mission in the Democratic Republic of the Congo (MONUSCO).[11] Initially, the operation's focus was on the ceasefire and disengagement of forces and maintenance of liaison with all parties involved with the civil war but then expanded to include the effective protection of civilians, humanitarian personnel and human rights defenders under imminent threat of physical violence. The presence of illegal armed groups, such as M23, has continued to be an issue that MONUSCO has monitored in recent years. In November 2012, M23 occupied the city of Goma, a provincial capital in eastern DRC in the North Kivu province, and other cities in eastern DRC and clashed with the Congolese national army.

During this time, the UN reported cases of sexual violence perpetrated by armed groups and members of the Congolese national army against women and children. While M23 eventually withdrew from the cities, the group's presence in the region continues. In February 2013, the UN reported that eastern DRC continues to be plagued by recurrent waves of conflict, chronic humanitarian crises, and serious human rights violations, including sexual and gender-based violence. The report added that contributing factors to the cycles of violence have been the continuing presence of Congolese and foreign armed groups taking advantage of security vacuums in the eastern part of the country, the illegal exploitation of resources, interference by neighboring countries, and the weak capacity of the national army and police to effectively protect civilians and the national territory and ensure law and order. In March 2013, the UN Secretary-General appointed a Special Envoy to the Great Lakes Region of Africa to support the implementation of the 11-nation "Peace, Security and Cooperation Framework for the Democratic Republic of the Congo and the Region" adopted in February 2013. According to the UN, the agreement seeks to end the recurring cycle of conflicts and crisis in the eastern DRC and to build peace. Additionally, on March 28, 2013, the UN Security Council authorized the deployment of an intervention brigade within the current peacekeeping operations in DRC to address imminent threats to peace and security. The objectives of the new force based in North Kivu province are to neutralize armed groups, reduce the threat they pose to state authority and civilian security, and make space for stabilization activities.

U.S. Government Response to Situation in the DRC

Congress has focused on issues related to the DRC for almost a decade. In 2006, Congress passed the Democratic Republic of Congo Relief, Security, and Democracy Promotion Act of 2006.[12] The act stated that it is the policy of the United States, among other things, to engage with governments working for peace and security throughout the DRC and hold accountable individuals, entities, and countries working to destabilize the government. In July 2010, Congress included several provisions in section 1502 of the Dodd-Frank Act related to conflict minerals in the DRC and adjoining countries.[13] Specifically, section 1502(a) of the Act states that "it is the sense of Congress that the exploitation and trade of conflict minerals originating in the [DRC] is helping to finance conflict characterized by extreme levels of violence in the eastern Democratic Republic of the Congo, particularly sexual- and gender-based violence, and contributing to an emergency humanitarian situation therein," warranting the provisions of Section 1502(b) of the Act. Section 1502(b) requires SEC, in consultation with State, to promulgate disclosure and reporting regulations regarding the use of conflict minerals from the DRC and adjoining countries. In November 2011, State and USAID, in collaboration with NGOs, industry, and other governments, launched the Public-Private Alliance for Responsible Minerals Trade (PPA) to support responsible supply chain solutions regarding conflict minerals from the DRC and neighboring countries. The PPA supports pilot programs, with the ultimate goal of producing scalable, self-sustaining systems, to demonstrate a fully traced and validated conflict-mineral supply chain in a way that is credible to companies, civil society, and government. According to USAID, in addition to the PPA, the U.S. government's contribution to the Responsible Minerals Trade Program in the DRC region has amounted to almost $19 million and includes activities focused on the protection of artisanal mining communities, institutional and human capacity building for responsible minerals trade, and capacity building in mining sector security, among other issues.

SEC's Final Conflict Minerals Rule

The SEC Commissioners adopted[14] the final conflict minerals rule on August 22, 2012, after a number of delays during the drafting process.[15] SEC reported that during its rule-making process it received more than 400 letters commenting on the draft rule.[16] As adopted, the final rule applies to any issuer that files reports with SEC under Section 13(a) or Section 15(d) of the Securities Exchange Act of 1934[17] (Securities Exchange Act) and uses conflict minerals that are necessary to the functionality or production of a product manufactured or contracted by that issuer to be manufactured. According to SEC, issuers that have a reporting obligation are domestic and foreign companies that offer shares publicly and file forms 10-K, 20-F, or 40-F with SEC. For the purposes of our report, we refer to those issuing companies affected by the rule as "SEC-reporting companies under the rule." (See app. II for more information on the steps a company needs to take to fulfill its reporting requirements.) Under the rule, such companies must file a disclosure report and conduct a "reasonable country of origin inquiry" to determine whether they must also file a conflict minerals report.[18]

Source: GAO analysis.

Figure 2. Timeline for SEC-Reporting Companies to Submit Conflict Minerals Disclosures to SEC.

Companies that are required to file a conflict minerals report must exercise due diligence on the source and chain of custody of their conflict minerals. The due diligence measures used by companies must conform to a nationally or internationally recognized due diligence framework, such as the due diligence guidance approved by OECD. If a company determines that its products are "DRC conflict-free" because they may have originated from the covered countries but did not finance or benefit armed groups, then the company must obtain an independent private sector audit and provide certification that it conducted an audit. If a company's products have not been found to be "DRC conflict-free," then the company must provide additional information in its conflict minerals report. For a temporary period—4 years for smaller reporting companies or 2 years for all other reporting companies—if a company is unable to determine whether the minerals in its products originated in the DRC or the adjoining countries or financed or benefited armed groups in those countries, then those products are considered "DRC conflict undeterminable" and no audit is required.[19] Under the rule, all companies will need to file their first disclosure report to SEC on May 31, 2014, which covers the 2013 calendar year, and on May 31 annually thereafter. Figure 2 shows the reporting time frames for SEC-reporting companies under the rule.

In October 2012, the U.S. Chamber of Commerce, the National Association of Manufacturers, and the Business Roundtable filed a lawsuit against SEC regarding the final conflict minerals rule.[20] In their petition, the two industry associations asked that the rule "be modified or set aside in whole or in part." The petitioners have asked the court to review, among other things, whether SEC's economic analysis is inadequate and whether SEC's interpretations of certain key terms in section 1502 of the Act are consistent with congressional intent.

Conflict Minerals Description and Supply Chain

The four conflict minerals covered by section 1502(b) of the Dodd-Frank Act are mined in various locations around the world. For example, tin is predominantly mined in China, Indonesia, Peru, and Bolivia, as well as in the DRC, while tantalum is reportedly predominantly mined in areas such as Australia, Brazil, and Canada. From 2006 through 2011, the majority of tungsten production—reportedly 77 to 87 percent of global production— was mined in China. Gold, however, is mined in many different countries, including the DRC. Our review of United States Geological Survey data on tantalum, tin, tungsten, and gold mined in the DRC showed that about 12 percent of the global tantalum supply and less than 1 percent of the global tungsten supply was mined in the DRC in 2011. About 3 percent of the global tin supply, and less than 1 percent of the global gold supply, was mined in the DRC in 2010. As we reported in our 2012 report, various industries, particularly in manufacturing, use these minerals in a wide variety of products and in varying amounts. For example, many industries use tin in the form of tin solder, which is used to join metal pieces together.[21] According to company representatives, tin is also found in food packaging, in steel coatings on automobile parts, and in some plastics. According to industry association and company representatives, the majority of tantalum is used to manufacture tantalum capacitors, which enable energy storage in electronic products such as cell phones and computers.[22] Tungsten is used in automobile manufacturing, drill bits and cutting tools, and other industrial manufacturing tools. It is also the primary component of filaments in light bulbs. In addition to its use as currency and in jewelry, gold is also used by other industries, such as the electronics industry.

A company's supply chain for products containing tin, tantalum, tungsten, and gold can be complex and can vary considerably in the way it operates, according to industry association and company representatives. Generally, however, the supply chain for companies using conflict minerals begins at the mine site, where tin, tantalum, and tungsten ore are extracted from the ground using mechanized or artisanal mining techniques. [23] Figure 3 provides additional information on a simplified supply chain for all four conflict minerals.

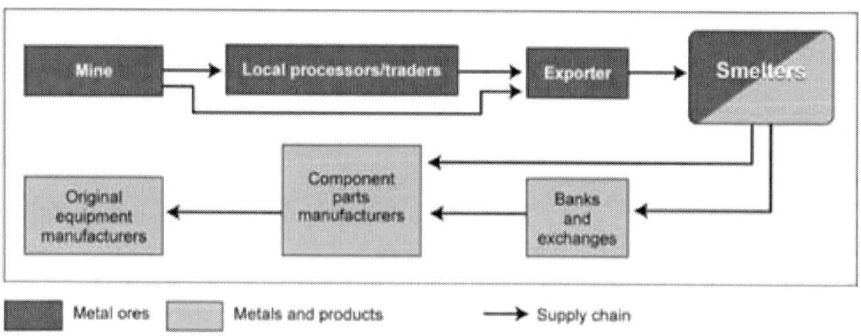

Source: GAO analysis.

Figure 3. Simplified Conflict Minerals Supply Chain.

For artisanal mining, the local processor or trader (which may be an individual or company) purchases minerals directly from the mine sites and typically processes or upgrades them before selling them to an exporter. The exporter may also purchase minerals directly from mine sites rather than going through a local processor or trader. Exporters may carry out further processing or upgrading before exporting materials to a smelter or refiner, where they are either converted into metals or purified into a higher-purity metal. Smelters primarily provide high-purity tin, tantalum, and tungsten directly to component parts manufacturers, although some sell high-purity metals through traders or exchanges.[24] Gold refiners typically sell high-purity gold to banks, for use as a store of value, or to international exchanges, where gold is bought and sold.[25] However, some gold refiners sell gold directly to manufacturers as well. Banks and traders may sell gold to manufacturers, including jewelry and component parts manufacturers. The component parts manufacturers construct individual parts—such as capacitors, engine parts, or clasps for necklaces—that they sell to original equipment manufacturers. The original equipment manufacturers complete the final assembly of a product and sell the final product to the consumer.

STAKEHOLDER-DEVELOPED INITIATIVES MAY FACILITATE COMPLIANCE WITH CONFLICT MINERALS RULE, BUT OTHER FACTORS MAY AFFECT THE RULE'S IMPACT ON REDUCING BENEFITS TO ARMED GROUPS

SEC's adoption of the final conflict minerals rule on August 22, 2012, has raised companies' awareness regarding conflict minerals and the due diligence necessary to identify whether conflict minerals may have benefited armed groups. Specifically, officials representing industry associations stated that the final conflict minerals rule has acted as an impetus for some of their members to start thinking about whether the rule impacts them and some have also started collecting information to comply with the rule. Officials stated that stakeholder-developed initiatives, such as in-region and global sourcing initiatives, may increase companies' assurance that conflict minerals they are using are not benefiting armed groups in the DRC and neighboring countries. However, constraining factors such as the lack of security, lack of infrastructure, and capacity constraints could undermine companies' ability to ensure conflict-free sourcing from the region.

Adoption of the Final Rule Has Raised Companies' Awareness Regarding Sourcing of Conflict Minerals

Since SEC issued the final conflict minerals rule pursuant to the Dodd Frank Act, companies have become more aware of the issues surrounding conflict minerals and have started to consider the source of materials used in products, given the requirement in the final rule for a company that uses tin, tantalum, tungsten, or gold to exercise due diligence on the source and chain of custody of its conflict minerals, if there is reason to believe that they may have originated in the DRC or an adjoining country.[26] According to some industry officials we interviewed, the final rule has helped resolve some uncertainties, such as the breadth of the industries covered, that existed before the promulgation of the rule. Numerous industry officials and representatives from international organizations and NGOs we interviewed have indicated that the creation and promulgation of the SEC rule has increased visibility into the issue of conflict minerals and raised awareness of the due diligence process, particularly for those companies that are not required to report under the rule but that may still be impacted indirectly by the rule.[27] Specifically, officials of industry associations representing member companies that use tin, tantalum, tungsten, or gold in their products stated that many companies are aware of the SEC rule, especially the larger companies that may file a disclosure report with SEC, and are working to start complying with the rule. Some smaller companies, which may not be required to report under the rule, may not be as aware of or familiar with the rule but are receiving information from industry associations on how the rule may impact them. For example, some officials from industry associations stated that they were putting together guidance documents that break down the SEC rule and had also sent questions to SEC seeking to clarify points in the rule. Agency officials stated that the SEC rule has raised visibility globally of conflict minerals. For example, State reported in February 2013 that the issuance of the SEC rule was a vital step in establishing a clear and harmonized global framework for responsible minerals trade from the DRC region. Furthermore, State indicated that the SEC rule has also shaped and influenced initiatives to create a conflict-free supply chain by the International Conference of the Great Lakes Region (ICGLR) and the governments of the DRC and Rwanda. We provide a more detailed discussion later in this report on the ways in which companies required to report under the rule, in order to comply, are interacting with companies not required to report under the rule.

Stakeholder-Developed Initiatives May Increase Companies' Assurance That Conflict Minerals They Use Are Not Benefiting DRC Armed Groups

Stakeholder-Developed Initiatives May Aid Companies' Due Diligence Efforts

Some agency officials we interviewed stated that stakeholder-developed initiatives focused on sourcing of minerals may enhance companies' ability to achieve the SEC rule's desired outcome of denying armed groups in the DRC benefits from conflict minerals. As mentioned in our 2012 report, stakeholder-developed initiatives—which include the development of guidance documents, audit protocols, and in-region sourcing—support efforts by companies reporting to SEC under the rule to (1) conduct due diligence of their conflict minerals supply chain, (2) identify the source of conflict minerals within their supply chain, and (3) responsibly source conflict minerals. These initiatives can be classified as in-region or global, and some are now being expanded.

In-Region Sourcing Initiatives Focus on Tracing of Conflict Minerals

In-region sourcing initiatives, as we reported in 2012, may support responsible sourcing of conflict minerals from Central Africa and the identification of specific mines of origin for those minerals. Regional sourcing initiatives in the DRC and neighboring countries focus on tracing minerals from the mine to the mineral smelter or refiner by supporting a bagging and tagging program or some type of traceability scheme. Examples of such initiatives include the ITRI Tin Supply Chain Initiative (iTSCi) and the Conflict-Free Tin Initiative (CFTI). (See app. III for more detailed information on these and selected other in-region sourcing initiatives). The iTSCi initiative was developed by a tin industry association known as ITRI. The initiative supports responsible sourcing of tin, tantalum, and tungsten from Central Africa and was launched in Rwanda in December 2010 and in the Katanga province of the DRC in March 2011. iTSCi expanded its activities in the Maniema province of the DRC in December 2012. We reported in 2012 that iTSCi is a traceability and due diligence program that creates auditable and verifiable chains of custody for tin, tantalum, and tungsten through (1) tagging and bagging of materials and the collection of tagging data and (2) regular incident reporting and continuous monitoring of mines and companies participating in the program. In October 2012, the Dutch government, with industry partners such as iTSCi, started the Conflict-Free Tin Initiative focused on conflict-free tin sourcing from South Kivu in the DRC, a region that is prone to insecurity and violence by illegal armed groups. This initiative is a traceability and due diligence mechanism that brings partners along the supply chain together, from mine to smelter to end-user. This also includes consumers as well as the DRC government and civil society[28] and uses the OECD due diligence guidance. According to an implementer, the progress of the initiative will depend on how the security situation in South Kivu develops.

While the in-region sourcing initiatives have focused on tin, tantalum, and tungsten to date, one of the most recent in-region initiatives in the DRC is through the Public-Private Alliance for Responsible Minerals Trade (PPA) and is focused on a pilot gold traceability scheme. In fall of 2012, Partnership Africa Canada began work on establishing an in-region gold traceability project partially funded by PPA. According to information from PPA, the project aims to create a traceable conflict-free mineral chain for artisanal gold from the eastern DRC, in the Orientale province, thus demonstrating the feasibility of creating artisanal gold chains with full traceability from mine site to gold refiner.

According to industry and agency officials, in-region sourcing programs can provide better economic incentive for miners to sell minerals that do not benefit armed groups. For example, iTSCi reported in February 2013 that the tin initiative in South Kivu had led to a number of immediate benefits for the local population, which depends on mining as its source of income. Specifically, the price paid to the miners for conflict-free minerals mined at the site had more than doubled. iTSCi further reported that the additional income had allowed the mining cooperatives to invest in basic equipment such as electricity generators and to improve productivity and working conditions. Additionally, some agency officials stated that in-region initiatives can help develop capacity in the DRC.

Global Sourcing Initiatives Affect Companies throughout the Supply Chain

We reported in 2012 that global sourcing initiatives may minimize the risk of minerals that have been exploited by illegal armed groups from entering the supply chain and support companies' efforts to identify the source of the conflict minerals across the supply chain around the world. (See app. III for more detailed information on selected global sourcing initiatives.) One such global initiative is the Conflict-Free Smelter Program, co-developed by the Global e-Sustainability Initiative (GeSI) and the Electronics Industry Citizenship Coalition (EICC). The Conflict-Free Smelter Program is a voluntary initiative in which an independent third party audits smelters' procurement activities—among other activities—and determines if the smelters demonstrated that the minerals they processed originated from conflict-free sources. Companies that can trace their conflict minerals supply chain back to compliant smelters or refiners can claim that the minerals in their products are from a smelter[29] whose processes reasonably assure conflict-free production.[30]

Industry experts we interviewed explained that if initiatives such as the Conflict-Free Smelter Program can result in smelters refining minerals that did not benefit armed groups, then companies can comply better with the SEC rule requirements and have more confidence in their supply chain sources. Specifically, one expert stated that companies that are conducting due diligence under the rule would not have to audit the smelter themselves, if the smelter has already been audited under the Conflict-Free Smelter Program. Agency officials, both in Washington, D.C., and in the DRC, reported that the Conflict Free Smelter Program seems to be a positive initiative since the more smelters are certified as conflict free, the more beneficial this will be for companies reporting under the SEC rule.

Overall, agency officials we interviewed stated that existing initiatives and traceability schemes on the ground in the DRC and neighboring countries have been yielding benefits of producing conflict-free minerals; however, according to these officials, more progress could be made in the responsible sourcing of conflict minerals from the region. Some industry experts also indicated that while progress has been made and the initiation or expansion of in-region sourcing initiatives is possible, factors such as the ones described below remain a concern.

Lack of Security, Inadequate Infrastructure, and Capacity Constraints in the DRC Could Affect Companies' Ability to Ensure Conflict-Free Sourcing of Minerals

Some agency officials as well as representatives we interviewed from NGOs, industry, and international organizations cited lack of security, inadequate infrastructure, and capacity constraints as factors that could affect the ability to expand on efforts to achieve conflict-free sourcing of minerals from the eastern DRC and thereby potentially contribute to armed groups benefiting from the conflict minerals trade. We also cited these same factors in our 2010 report and pointed out that these factors posed challenges to tracking the mines of origin for minerals artisanally mined in eastern DRC.[31] While officials we spoke to for this report discussed these factors in the context of the SEC rule, these factors are pre-existing regional challenges that pre-date both the Dodd Frank Act and the SEC conflict minerals rule.[32]

Lack of Security in Eastern DRC Could Hamper Sourcing Initiatives

Officials cited the lack of security, including weak governance, as a factor that could impact responsible sourcing from the DRC. The UN reported that the DRC government has been unable to exercise authority in eastern DRC, which has become more evident as illegal armed groups clashed in the Kivu provinces late in 2012. State also reported that lack of security has prevented the export of conflict-free minerals from certain areas in eastern DRC. Industry and NGO officials who work on the ground in the DRC pointed out that the threat from illegal armed groups poses a challenge to the conflict-free minerals initiatives operating in eastern DRC and the neighboring provinces. Although the mining sites are constantly monitored, the monitoring activities could be suspended at any time as the security situation evolves. For example, an NGO reported that tagging was suspended for days in July 2012 at an iTSCi site in the Katanga province because of the movement of armed groups in the vicinity of the mine sites; however, there were no reported cases of armed groups successfully taking control of the sites or directly exploiting minerals to fund activities.

In-region sourcing initiatives have operated in areas that have been vetted by various stakeholders and have the support of government and civil society actors. According to the UN Group of Experts on the Democratic Republic of the Congo (UNGoE), the security situation at tin, tantalum, and tungsten mining sites has improved and the trade of these minerals has become a much less important source of financing for armed groups. However, the UNGoE reported a "genuine risk that military actors would move their rackets to mining activities that were not closely supervised." They further reported that the gold trade is linked to armed groups and criminal networks in the Congolese armed forces. According to the UNGoE, lack of security at gold mining sites throughout eastern DRC remained widespread. Agency officials emphasized that armed groups still existed in the DRC despite the initiatives in place and would seek control of any significant revenue-producing activity in the region.

Some industry officials cited concerns about sourcing from the DRC, even through the in-region sourcing initiatives, because of the potential impact on brand reputation and financial risk. For example, a representative of a smelter indicated that if the company purchased minerals from a mine that is part of a traceability scheme that is deemed conflict-free but then illegal armed groups infiltrated and compromised the mine in the future, the company would not be able to say with certainty that the minerals it had purchased were conflict-free.

Limited Infrastructure Could Hamper Access to Mines

Officials cited limited infrastructure as a factor that could affect the creation or expansion of in-region sourcing initiatives. Officials from UNGoE and industry representatives we interviewed noted a lack of infrastructure in place that would enable companies to set up or expand operations in the DRC. Limited transportation and poor roads in eastern DRC also make it difficult to get to mine sites. For example, an agency official in the DRC commented that mines may be a day's walk from a main road. Also, an NGO reported that in selecting a potential pilot site for a traceability scheme, accessibility to the site by road was a key criterion and would involve using off-road vehicles due to the significant deterioration of roads leading to the mine. Moreover, according to an NGO representative, the remoteness of mines also makes it difficult for DRC mine officials to validate mines and ensure that the mines have not been compromised by armed groups. Furthermore, State officials indicated that the lack of infrastructure prevents trade initiatives from developing economies of scale and expanding.

Lack of Capacity Could Impact due Diligence

Officials we interviewed cited the lack of technical, economic, and political capacity as another factor that may affect the creation or expansion of in-region sourcing initiatives focused on responsible sourcing in the DRC and neighboring countries. In 2013, the OECD reported that while the understanding of responsible sourcing is "high for those actors in the DRC and Rwanda who have participated in such initiatives, the same is not true for state agents" in the country.[33] The OECD report also pointed out that Ugandan and Burundian government officials and other entities lack technical understanding of due diligence requirements. Some NGO officials stated that lack of capacity can impact the due diligence process in the supply chain, especially if the numbers of trained mining agents is insufficient. For example, some agency officials and an NGO reported that the DRC does not have enough mine agents to certify the mines, of which there may be over 2,000 in eastern DRC alone, or even to negotiate and manage mining contracts. Moreover, an NGO official stated that mines need to be reinspected every 6 to 12 months in order to ensure proper due diligence in accordance with OECD and ICGLR guidance; however, the NGO official stated that the DRC government does not have the capacity to inspect at such frequency.

Some agency officials and officials we interviewed from industry, NGOs, and international organizations also commented that the DRC government lacks capacity to mitigate corruption and smuggling. The lack of capacity can impact due diligence and can contribute to illegal minerals trade and cross-border smuggling. For example, the UN reported that illegal trade of minerals undermines the exercise of due diligence in the DRC and affects the credibility of due diligence-based certification and traceability systems. [34] According to some industry experts, mining agents may not be properly compensated, due to the lack of governance in eastern DRC, and may look for other ways to earn money, which could involve colluding with illegal armed groups.

With regard to smuggling, the OECD reported that as long as there are no traceability or certification schemes in place that cover the whole region, and most notably, the Kivu provinces, Uganda, and Burundi, smuggling and contamination of clean materials will continue to pose a threat to formalization of the artisanal mining sector and due diligence initiatives. According to a 2012 UNGoE report, several tons of gold worth hundreds of millions of dollars are smuggled from the eastern DRC through neighboring countries, where it is ultimately smelted and sold to jewelers in markets, such as the United Arab Emirates. Representatives from some industry associations that we interviewed stated that armed groups and criminal elements have shifted efforts to gold mines because it is relatively easy to smuggle gold because of its size. Furthermore, gold's high value in the market makes it more viable for smuggling than tin, tantalum, and tungsten.

Many Companies Not Required to Report to SEC under the Conflict Minerals Rule Will Likely Be Affected; Limited Aggregated Information about Them Exists

Even companies that are not required to file disclosures under SEC's conflict minerals rule will likely be affected by the rule. These companies may supply components or parts that contain conflict minerals to companies reporting to SEC under the rule and may be asked by such companies to provide information specifying the origin of the minerals. Aside from the supply chain relationship, while information is publicly available about some smelters and refiners, there is little aggregated information available about companies that do not report to SEC under the rule but may trade in conflict minerals.

Companies Not Required to Report to SEC under the Rule May Provide Information on the Origin of Conflict Minerals in Their Products to Companies That Will Report to SEC under the Rule

Companies Not Required to Report to SEC under the Rule May Supply Products That May Contain Conflict Minerals to SEC-Reporting Companies under the Rule

Companies that are not required to report to SEC under the rule may supply products that contain conflict minerals to SEC-reporting companies under the rule. SEC relied on estimates provided by a commentator indicating that 278,000 suppliers—most of which would be companies that would not report to SEC under the rule—could be indirectly impacted by the rule. Moreover, the release contains an estimate that each of the nearly 6,000 companies that could be directly impacted by the rule has roughly 1,000 first-tier suppliers, on average.[35] These suppliers, including first-tier suppliers, could provide products that contain conflict minerals to companies required to report to SEC under the rule. Examples of these products include tin solder for joining metal, tantalum capacitors for storing energy in cellular phones, tungsten carbide for hardened cutting tools, or gold plating for wires to increase durability and resistance to corrosion. The first-tier supplier has a direct commercial relationship with the original equipment manufacturer, meaning the first-tier supplier sells materials or component parts, which have been aggregated by suppliers throughout the supply chain, to the original equipment manufacturer for final assembly. According to an industry official, in general, component parts manufacturers construct individual parts—such as capacitors, engine parts, circuit boards, and other components—and assemble them into more complex components. Using an electronics company as a model, processed metals move through several suppliers that manufacture component parts after the smelter—first to circuit board and computer chip manufacturers, then to cellular phone and other electronics manufacturers, and finally to the brand-name electronics company, which is the original equipment manufacturer that manufactures products recognizable to the consumer, such as cellular phones, tablets, and laptop computers. Beyond the first-tier supplier, there are tier 2-, 3-, 4-, or higher-tiered suppliers that, beginning with the raw materials from the smelter or refiner, manufacture component parts that are assembled into more complex component parts as they move from higher- to lower-tiered suppliers in the supply chain, to the first-tier supplier, and finally to the original equipment manufacturer.[37] See figure 4 for a simplified version of the supply chain, and the tiered structure of suppliers.

Estimate of the Number of Companies Required to Report to SEC under the Rule

SEC estimates that 5,994 reporting issuers (primarily companies that issue stock publicly and are required to report to SEC) will be affected by the rule and will need to determine if their products contain conflict minerals. According to SEC, reporting issuers are domestic and foreign companies that file forms 10-K, 20-F, and 40-F under the Securities Exchange Act.[36] According to SEC and industry officials, these companies vary in size and revenue, but in general, tend to be larger, mature companies that can have a diverse product line; their revenues can range from the millions of dollars to hundreds of billions; they are domestic and foreign companies; and they may have operations in several countries. According to SEC and industry officials, some reporting issuers may sell products to consumers.

Source: GAO analysis of SEC information.

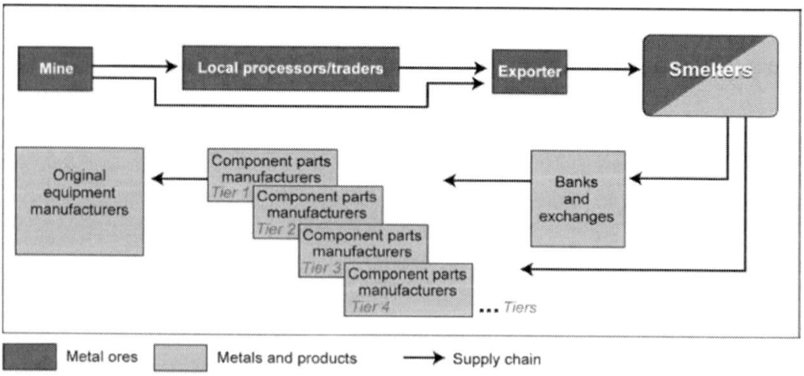

Source: GAO analysis.

Figure 4. Simplified Conflict Minerals Supply Chain Showing Supplier Tiers.

While many companies will likely be directly or indirectly impacted by the rule, some companies that use conflict minerals may not be, partly because (1) the companies are not issuers that are required to file with SEC under the Securities Exchange Act, and (2) these same companies potentially do not sell components or parts to a company that will be required to report to SEC under the rule. Industry and consulting firm representatives have differing views on the number of companies that purchase conflict minerals from the DRC and adjoining countries but may not be impacted by the rule.

SEC-Reporting Companies under the Rule May Request Information on the Origin of Conflict Minerals in Products They Receive from Their Suppliers

Suppliers that provide products that may contain conflict minerals to companies required to report to SEC under the rule may provide information on the minerals' origins to those reporting companies that request it. The SEC release does not specify the steps and outcomes for the reasonable country of origin inquiry, and indicates that such a determination depends on each issuer's facts and circumstances. However, in conducting a country of origin inquiry, issuers may inquire of their suppliers the origin of any conflict minerals in the products. According to the release, the issuer's inquiry must be reasonably designed to determine whether any of its conflict minerals originated in the DRC and adjoining countries, and must be performed in good faith. If, after this inquiry, the issuer has a reason to believe that its conflict minerals may have originated in the DRC and adjoining countries, the issuer proceeds to exercising due diligence. Industry associations such as the EICC and GeSI have created templates for companies to use when contacting suppliers to inquire about the types and origins of conflict minerals in a given product.[38] For example, companies required to report under the rule could submit the inquiries to their first-tier suppliers. Those suppliers could either provide the reporting company with sufficient information or initiate the inquiry process up the supply chain, such as by distributing the inquiries to suppliers at the next tier—tier 2 suppliers. The tier 2 suppliers could inquire up the supply chain to additional suppliers, until the inquiries arrive at the smelter. Smelters then could provide the suppliers with information about the origin of the conflict minerals. Figure 5 illustrates the flow of information up the supply chain.

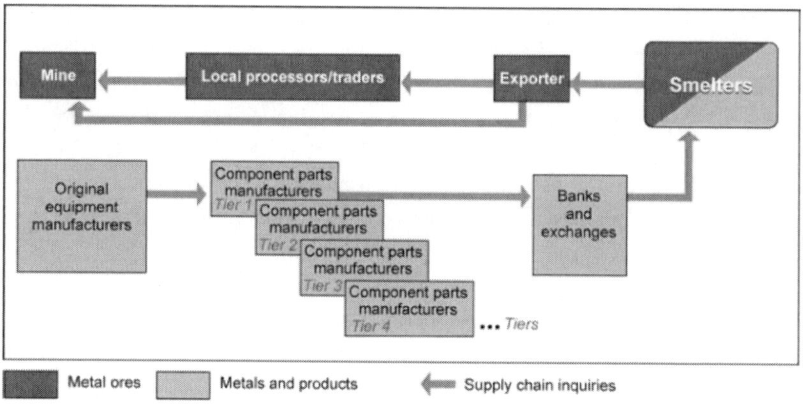

Source: GAO analysis.

Figure 5. Flow of Supply Chain Inquiries from the Original Equipment Manufacturers to the Mine.

As discussed earlier, smelters have various means to preclude untraced minerals from entering their supply, such as participation in the iTSCi initiative and the Conflict Free Smelters Program. According to smelting industry representatives, these initiatives and certifications have reduced the burden of responding to the multiple amounts of inquiries many smelters have already received from suppliers.

Officials from consulting firms and industry associations that we spoke with told us that many companies that will respond under the rule have started contacting their first-tier suppliers and providing them with country of origin inquiries. According to these officials, several of these companies that have submitted inquiries to their suppliers have experienced challenges, which include identifying suppliers beyond the first-tier suppliers, because for original equipment manufacturers, suppliers beyond the first tier are less visible. As discussed earlier, original equipment manufacturers purchase component parts primarily from their first-tier suppliers and do not have direct commercial relationships with suppliers in higher tiers of the supply chain. According to industry representatives and agency officials, these challenges may impact how companies file under the rule. For example, as previously discussed, the SEC rule allows companies to disclose their products as "DRC conflict undeterminable." This provision allows companies to state that the source of the conflict minerals in their products, and the likelihood that the conflict minerals benefited or financed armed groups from the DRC and adjoining countries, could not be determined after having conducted due diligence to obtain that information from their suppliers. For the reporting period beginning January 1, 2013, companies may use this provision for 4 years for smaller reporting companies or 2 years for all other reporting companies. Although the number of companies required to report under the rule that may utilize the "DRC conflict undeterminable" provision is unknown, SEC officials and representatives of industry associations and consulting firms anticipate that many companies required to report under the rule will utilize the provision based on the results of their due diligence efforts.

Representatives from industry and consulting firms that we interviewed stated that the purchasing power of issuing companies under the rule may influence their suppliers to provide information on the source of any conflict minerals in their products when requested. According to an industry representative, since companies that report to SEC under the rule tend to be large, mature corporations with great purchasing power in their respective industries, it would be difficult for suppliers to ignore their request for information on the origin of conflict minerals in products the suppliers provide to them. For example, jewelry industry representatives told us that they have advised their members, which are primarily small, independent jewelry companies not required to report to SEC under the rule, to respond to any requests from customers seeking information on the origin of conflict minerals in products they supply, because the risk of not responding could result in a loss of business for those companies.

Information Is Publicly Available about Smelters and Refiners, and the Types of Conflict Minerals They Use

Some information is publicly available about smelters and refiners, and their involvement in the conflict minerals supply chain. According to SEC officials, while smelters and refiners are not exempted from the SEC rule, most of these suppliers will likely not be required to report to SEC under the rule because of their filing status.[39] Smelter and refiners are considered the choke-point of the conflict minerals supply chain, as previously discussed, and comprise a small portion of the overall number of suppliers in the conflict minerals supply chain that may be impacted by the SEC rule. While it is not possible to determine the universe of suppliers that would not be required to report under the rule, smelters and refiners are a more identifiable population for which there is some aggregated information, such as the types of conflict minerals they use, and their location.[40] We found the following information on smelters and refiners:

- Smelters and refiners constitute a small but important portion of suppliers that likely will not file a conflict minerals report under the SEC rule. Organizations have estimated the number of smelters and refiners around the world to be nearly 500; however, the actual number of smelters and refiners of conflict minerals is unknown. We aggregated publicly available information on smelters and refiners from lists compiled by the EICC and GeSI, the OECD, and the London Bullion Market Association (LBMA), which provided information on 278 smelters and refiners. As we have previously discussed, roughly 278,000 suppliers could be affected by the rule, based on the estimate provided to SEC. Of the 278 smelters of tin, tantalum, tungsten, and refiners of gold that we were able to identify, the majority (271) of these companies would likely not be required to report under the rule.
- Over half of the smelters and refiners of the conflict minerals we identified were located in three countries. Of the 278 smelters and refiners of tin, tantalum, tungsten, and gold that we were able to identify, more than half (156) were located in three countries: China (82), Japan (39), and Indonesia (35). According to industry representatives, participation in due diligence efforts of smelters and refiners from these countries, particularly China and Indonesia, is critical in assisting companies with fulfilling the reporting requirements of the SEC rule. Several organizations, including an NGO, and representatives from government and industry, are conducting outreach to smelters to provide information on the SEC rule in an effort to increase participation from smelters and refiners from these countries. For more information on the location of smelters and refiners we identified in our analysis, see figure 6.
- *Many smelters and refiners of conflict minerals in our analysis processed tin.* Of the 278 smelters and refiners in our analysis, we were able to identify 113 that processed tin, followed by gold (78), tungsten (54), and tantalum (33). Furthermore, over half (64 of 113) of the tin smelters in our analysis were located in China (30) or Indonesia (34). In addition, over 67 percent of global production comes from mines in China and Indonesia, according to U.S. Geological Survey data. Tin and its derivatives have wide applications and are used in manufacturing a variety of products, including tin soldering for joining pipes, coatings for steel containers, and a wide range of tin chemical applications, according to the U.S. Geological

Survey. According to industry and consulting firm representatives, around 12 to 15 smelters process nearly 80 percent of the world's tin.

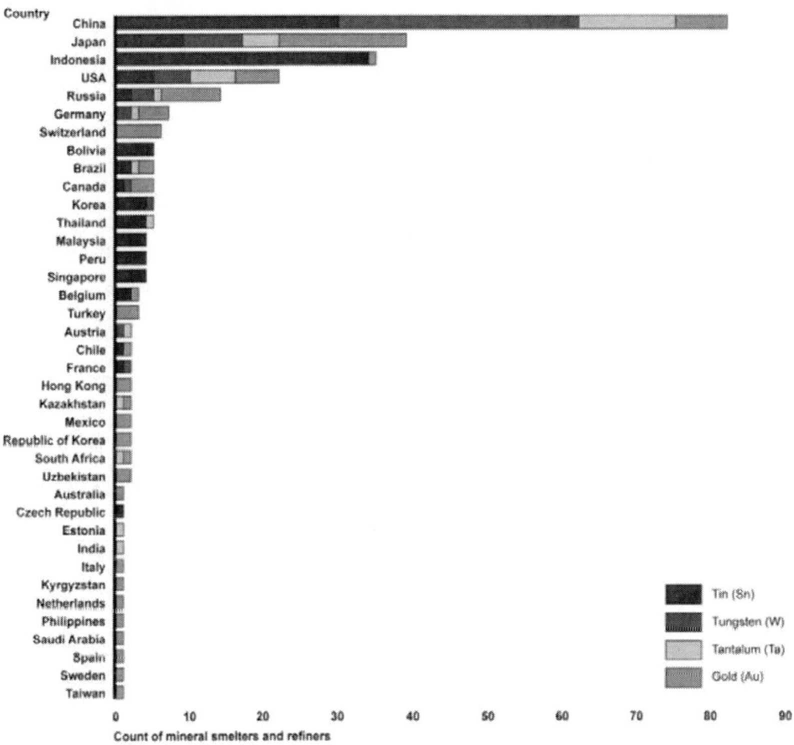

Source: GAO analysis of smelter information provided by the EICC-GeSI, OECD, and the LBMA.

Note: Eight smelters and refiners in our analysis processed more than one conflict mineral, resulting in duplicative counting in 21 cases. In addition, some smelters and refiners have operations in several countries.

Figure 6. Number of Smelters and Refiners of Conflict Minerals We Identified, by Country.

- *Most smelters and refiners in our analysis did not have a conflict minerals policy publicly available.* Of the 278 smelters and refiners we were able to identify, 63 had a conflict minerals policy publicly available on their website. Of the 63 smelters with a conflict minerals policy publicly available, 26 smelters had successfully completed a Conflict Free Smelter Program audit and were designated as "conflict-free" by the EICC and GeSI,[41] while several had reportedly followed some sort of due diligence, such as the OECD Due Diligence Guidance or the LMBA Responsible Gold Guidance.[42] Other smelters in our analysis had posted policies on their website stating that the company only sources conflict minerals outside of the conflict areas of the DRC and adjoining countries. We were unable to identify a website for 86 of the 278 smelters in our analysis, and 129 of the 278 smelters had no conflict minerals policy publicly available on their website.[43]

Some information is publicly available on companies that use conflict minerals but are not required to report under the rule, as in the case of many smelters and refiners we were able to identify. However, data for the universe of these companies are limited. Specifically, based on our analysis, aggregated data on the types of conflict minerals in the products manufactured by these companies as well as information on how such companies source their conflict minerals are not available, except for a few companies. For example, several of these companies provide information publicly about their continued participation in initiatives that source conflict minerals from the DRC, and have agreed to purchase conflict minerals, such as tin and tantalum, from closely monitored sources through initiatives such as iTSCi and the Solutions for Hope, as previously mentioned.[44] However, according to agency and international organization officials, in some instances buyers from small firms, mainly from East Asia, are on the ground in the DRC and adjoining countries, and continue to purchase untraced minerals as well as minerals that have been smuggled out of the DRC into adjoining countries. In addition, according to an industry representative, it may be difficult to identify information on these companies because they tend to be small and serve very specific markets.

Little Additional Information on the Rate of Sexual Violence in Eastern DRC and Neighboring Countries Has Become Available since GAO's 2012 Report

Since our 2012 report, one population-based survey providing data on the rate of sexual violence has been published in Uganda, and one is under way in the DRC; during the same period, no similar surveys have been conducted in

Rwanda or Burundi. We also found some additional case file data available on sexual violence for all four countries. However, as we reported in 2011, case file data on sexual violence are not suitable for estimating a rate of sexual violence.

Since GAO's 2012 Report, One Population-Based Survey on the Rate of Sexual Violence Has Been Published in Uganda and One Is Planned for DRC

We found that one new population-based survey on the rate of sexual violence has been conducted since our 2012 report[45]—the 2011 Uganda Demographic and Health Survey (DHS), published in August 2012. According to the survey, "28 percent of women and 9 percent of men age 15-49 report that they have experienced sexual violence at least once in their lifetime." These national estimates are based on a random sample. Since we first reported on sexual violence in our 2011 report, we have identified six other population-based surveys that provided data on the rate of sexual violence in these countries.

Surveys Are More Appropriate for Estimating a Rate of Sexual Violence

In our 2011 report on sexual violence, we discussed two sources of data on sexual violence in eastern DRC and neighboring countries—population-based surveys and case files—and concluded that population-based surveys are more appropriate for estimating a rate of sexual violence. Case file data have shortcomings and biases that significantly limit their utility for estimating the rate of sexual violence. For example, case file data are not generated from a random sample; are reliant on victims seeking services to be counted, although some victims may lack access to service; and allow for the potential double counting of the same sexual violence incident in the case file data collected.

Source: GAO.

In reviewing whether there had been updates to any of the previous surveys conducted, we found that the authors of the McGill study, a population-based survey conducted in eastern DRC that was highlighted in our

2011 report, had no plans to conduct a follow-up survey. We found that fieldwork for a DHS for the DRC is expected to launch in August 2013, with data expected around September 2014.

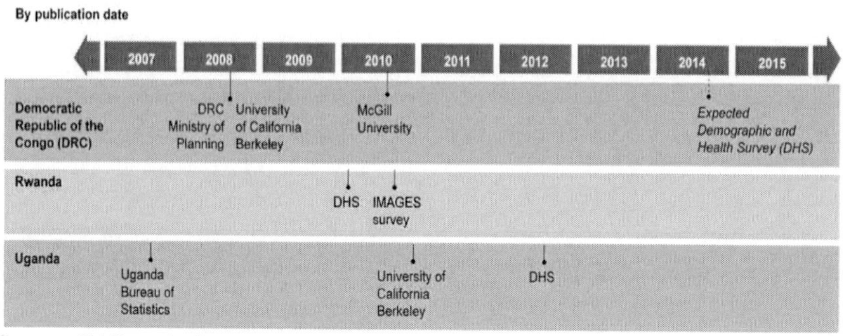

Source: GAO analysis.

Notes: For our analysis of the quality of the surveys, refer to our 2011 report (GAO-11-702). For DHS surveys covered after our 2011 report, we found that these surveys are weighted, present confidence intervals, and adhere to other survey standards. During our search for studies published since our 2012 report, we discovered the IMAGES survey, which we had not previously reported on in either our 2011 or 2012 report.

Figure 7. Population-Based Surveys That Estimate the Rate of Sexual Violence in Eastern DRC, Rwanda, and Uganda.

We also found a team of two organizations that released estimates in 2010 based on survey data of sexual violence in Rwanda. Sonke Gender Justice Network and Promundo-US conducted a probability cluster sample in 2010 as part of its IMAGES survey and found that "57 percent of women reported having experienced gender-based violence committed by a partner" and "17 percent of men experienced sexual violence when they were a child." However, this survey was not weighted to reflect unequal probabilities of selection, and it does not contain confidence intervals. Therefore, we are not able to assess the accuracy or precision of the estimates.

Some Additional Case File Data Has Become Available on Sexual Violence since GAO's 2012 Report

Following up on our 2011 and 2012 reports, we asked U.S. and UN agencies as well as researchers and NGOs if they had any updated case file data. In April 2013, State submitted its annual country reports on human rights practices to Congress, which provided case file information pertaining to sexual violence in the DRC and neighboring countries. The 2012 *Department of State Human Rights Reports* reported the following:

- In DRC, the Ministry of Gender reported 10,037 cases of sexual- and gender-based violence in 2011 in eastern DRC.
- In Rwanda, prosecutors reported that they investigated 351 cases of rape in 2012. Of those 351 cases, 109 were filed in courts, 143 were dropped, and 99 were pending investigation.
- In Uganda, 520 cases of rape were reported in 2011, of which 269 were tried.
- In Burundi, Centre Seruka, a clinic for rape victims averaged 121 rape cases per month between January and September 2012.

Various UN entities reported other case file data. In March 2013, the UN Secretary-General reported that 764 people had become victims of sexual violence in eastern DRC from December 2011 and through November 2012.[46] In May 2013, the UN Joint Human Rights Office reported that an armed group committed 135 cases of sexual violence from November 20, 2012 through November 30, 2012. Furthermore, in December 2012, the UN Office for the Coordination of Humanitarian Affairs reported 70 rapes in Minova, a town in eastern DRC, from November 30 through December 4, 2012. Because case file data are not aggregated across various sources and the extent to which various reports overlap is unclear, it is difficult to obtain complete data on case files or even a sense of magnitude. One shortcoming of both case file data and surveys is that time frames, locales, and definitions of sexual violence are not consistent across data collection operations. As we reported in 2011, case file data on sexual violence are not suitable for estimating a rate of sexual violence because case file data are not based on a random sample and the results of analyzing these data are not generalizable.

Agency Comments and Our Evaluation

We provided a draft of this report to SEC, State, and USAID, for their review and comment. SEC, State, and USAID provided technical comments, which we incorporated in this report as appropriate.

We also provided relevant portions of the draft of this report to relevant external stakeholders for their technical comment. We received technical comments from some of these stakeholders, which we incorporated throughout this report as appropriate.

Lawrance L. Evans, Jr.
Director,
International Affairs and Trade

APPENDIX I: OBJECTIVES, SCOPE, AND METHODOLOGY

To describe factors that may impact whether the Securities and Exchange Commission's (SEC) conflict minerals rule denies armed groups in the Democratic Republic of the Congo (DRC) and adjoining countries benefits from conflict minerals, we interviewed officials from SEC, the Department of State (State), and the United States Agency for International Development (USAID), as well as representatives from international organizations, nongovernmental organizations (NGO), industry associations, consulting firms, and smelters and refiners of tin, tantalum, tungsten, and gold to get their views on the final SEC rule as well as any impacting factors. We chose the experts and stakeholders we interviewed to capture a range of perspectives about the types of minerals traded and because we had established contacts with these entities on our last review. In addition, some of the stakeholders we talked to have been working on the ground in the DRC. These experts and stakeholders constitute a nongeneralizable sample. The information gathered cannot be generalized and cannot be used to infer views of other experts or stakeholders cognizant of conflict minerals issues.

We reviewed Section 1502 of the Dodd-Frank Wall Street Reform and Consumer Protection Act (Pub. L. No. 111-203); reports and other documents from relevant U.S. agencies, such as SEC's final conflict minerals rule, press releases, and statements; reports issued by the UN Group of Experts on the Democratic Republic of the Congo (UNGoE) and the Organisation for Economic Cooperation and Development (OECD); as well as documents and reports from industry associations and NGOs. We did not travel to the DRC or speak with government officials in the DRC but obtained perspectives on issues from some stakeholders who operate in the DRC.

To identify and describe available information about entities that use conflict minerals and do not report to SEC under the rule, we interviewed officials from SEC, State, and USAID, as well as representatives from international organizations, NGOs, industry associations, consulting firms, and smelters and refiners of tin, tantalum, tungsten, and gold to get their views on the extent to which information is publicly available on companies that are not required to report under the rule that may use conflict minerals in their products, and the source of the conflict minerals. We reviewed and analyzed reports and other documents from organizations such as the OECD, London Bullion Market Association (LBMA), and the Electronics Industry Citizenship Coalition and the Global e-Sustainability Initiative (EICC and GeSI), as well as documents and reports from industry associations and NGOs. In addition, we conducted searches in the Nexis database using selected Standard Industrial Classification (SIC) codes listed under the Manufacturing division. Overall, there were 20 subcategories under the Manufacturing division of SIC codes, which include subcategories such as Tobacco Products; Paper and Allied Products; and Electronic and Other Electrical Equipment and Components, Except Computer Equipment. We selected SIC codes under the Manufacturing division for industries that have a higher likelihood of using conflict minerals in their product, such as Electronic and Other Electrical Equipment and Components, Except Computer Equipment. Through the database analysis, we were able to determine the filing status, location, revenue, and industry classification of the companies. We were unable to determine the types of products the companies produced, and the types of conflict minerals potentially used in the manufacturing process of their products. Because SIC codes do not indicate specific products, we were unable to use the Nexis data to develop an aggregate description of entities that use conflict minerals but do not report to SEC under the rule.

We compiled a list of smelters and refiners—which are a smaller universe of companies that are primarily not required to report under the rule— from the EICC and GeSI's Conflict Minerals Reporting Template and Dashboard, OECD's *Final Downstream Report On One-Year Pilot Implementation of the Supplement on Tin, Tantalum, and Tungsten*, and the LBMA's Good Delivery List. The data were current as of March 15, 2013. We selected these smelters and refiners because information is publicly available on the types of minerals these smelters and refiners process; however, we did not conduct an audit to verify how these entities sourced materials for processing. To compile our list of smelters and refiners, we reviewed and compared the lists from each source to identify and delete duplicate smelters and refiners. Additional duplicates were identified and deleted as a result of Internet searches using the names of the smelters and refiners.

While we made efforts to eliminate duplicate information where possible, smelters and refiners may be listed under different names, and therefore, some duplicate information may exist in the data. Information included in the list of smelters from the EICC and GeSI, OECD, and LMBA included (1) the location of the smelter or refiner, (2) the types of minerals smelted or refined, and (3) the due diligence guidance reportedly followed, in some cases. We identified 278 smelters and refiners of tin, tantalum, tungsten, and gold, and analyzed any publicly available information—mainly information posted on the companies' websites or information provided on the websites of organizations such as the EICC and GeSI or the LBMA—on their practices and policies for sourcing conflict minerals. This analysis included examining websites of 192 of 278 smelters and refiners to identify the types of due diligence guidance they reported to use to determine the country of origin of their conflict minerals sources. We were unable to identify the websites for 86 smelters or refiners on our list, which could have been the result of a smelter not possessing a website, or differing translations of company names from foreign characters—such as Chinese script or the Cyrillic alphabet—to the Roman alphabet. Additional limitations included our sample of 278 smelters and refiners, as organizations have estimated that the number of smelters and refiners is nearly 500, particularly if smaller smelters and refiners that process ores into metals at the mine site are included. These smelters and refiners, or secondary smelters, often have small operations and may not have a website, according to an industry representative.

Furthermore, the number of gold refiners could potentially be much larger, considering that little equipment and space is required to refine gold, depending on the quality; and gold can be refined at the mine site. The 278 smelters and refiners we were able to identify may not be representative of others, and the information we report about these 278 cannot be generalized to other smelters and refiners of tin, tantalum, tungsten, and gold.

In response to a mandate in the Dodd-Frank Wall Street Reform and Consumer Protection Act that GAO submit an annual report that assesses the rate of sexual violence in war-torn areas of the DRC and adjoining countries, we identified and assessed any additional published information available on sexual violence in war-torn eastern DRC, as well as three neighboring countries that border eastern DRC—Rwanda, Uganda, and Burundi—since our 2012 report on sexual violence in these areas.[1]

During the course of our review, we interviewed officials from State and USAID and interviewed NGO representatives and researchers to discuss the collection of sexual violence-related data—including population-based surveys and case file data—in the DRC and adjoining countries. Specifically, we followed up with researchers and representatives from those groups we interviewed for our prior review on sexual violence rates in eastern DRC and neighboring countries, including officials from the United Nations Population Fund, United Nations High Commissioner for Refugees, United Nations Special Representative to the Secretary-General on Sexual Violence in Conflict; and representatives from the Harvard Humanitarian Initiative and others. We also conducted Internet literature searches to identify new academic articles containing any additional data on sexual violence.

We conducted this performance audit from November 2012 to July 2013 in accordance with generally accepted government auditing standards. Those standards require that we plan and perform the audit to obtain sufficient, appropriate evidence to provide a reasonable basis for our findings and conclusions based on our audit objectives. We believe that the evidence obtained provides a reasonable basis for our findings and conclusions based on our audit objectives.

APPENDIX II: SEC'S FLOWCHART SUMMARY OF THE DISCLOSURE PROCESS FOR THE FINAL CONFLICT MINERALS RULE

The Securities and Exchange Commission (SEC) issued a flowchart summary of the final rule to guide SEC-reporting companies affected by the rule through the disclosure process (see figure 8). In general, the process shows that an SEC-reporting company needs to (1) determine whether its manufactured products contain conflict minerals; (2) determine whether conflict minerals are necessary to the product and, if so, whether the conflict minerals originated in the DRC or an adjoining country; and (3) possibly conduct due diligence and potentially provide a Conflict Minerals Report.

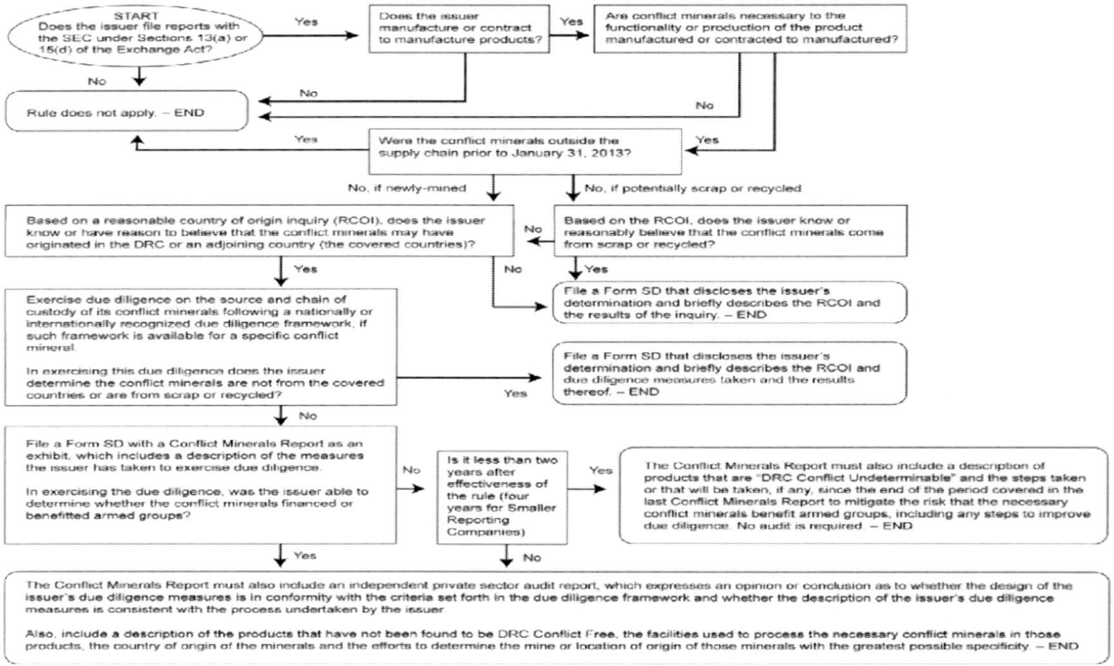

Source: SEC.

Note: SEC indicated that the flowchart is intended to be used as a guide and that issuers should refer to the text in the rule for a more comprehensive description of the rule's requirements.

Figure 8. SEC's Flowchart Summary of the Final Conflict Minerals Rule.

APPENDIX III: UPDATES OF GLOBAL AND IN-REGION SOURCING INITIATIVES

In our 2012 report, we discussed a number of initiatives that various stakeholders developed and implemented that may help companies reporting to the Securities and Exchange Commission (SEC) and their suppliers comply with SEC's conflict minerals disclosure rule. For this report, we updated information pertaining to some of these global and in-region sourcing initiatives.[1]

Global Sourcing Initiatives

OECD due Diligence Guidance

The Organisation for Economic Co-operation and Development (OECD) adopted the OECD Due Diligence Guidance for Responsible Supply Chains of Minerals from Conflict-Affected and High-Risk Areas (hereafter referred to as OECD Due Diligence Guidance) to promote accountability and transparency in conflict minerals supply chains.[2] The OECD Due Diligence Guidance and the corresponding supplements provide detailed guidance for companies operating in and sourcing minerals from conflict areas. In addition to the basic framework, OECD developed two supplements—one on tin, tantalum, and tungsten and the other on gold— to provide companies with specific guidance relevant to the conflict minerals supply chains. To increase awareness of and to develop emerging practices for implementing the OECD Due Diligence Guidance and the supplement on tin, tantalum, and tungsten, OECD conducted implementation pilot projects. In January 2013, OECD issued the final downstream report, which focuses on how companies implement due diligence in the supply chains of tin, tantalum, and tungsten, and the final upstream report, which provides an overall assessment of the progress and initial impact of due diligence in the tin, tantalum, and tungsten upstream supply chain.[3]

Conflict-Free Smelter Program

The Conflict-Free Smelter Program is a voluntary program in which smelters undergo an independent third party audit, in accordance with the OECD Due Diligence Guidance, to verify the origin of minerals processed at their facilities. The EICC and GeSI have also developed audit protocols for the program in consultation with a number of stakeholders—including NGOs, smelters, component manufacturers, original equipment manufacturers, and industry associations within and outside the electronics industry—to ensure wide-spread support for the program. In December 2010, the first tantalum smelter was certified conflict-free through the program after successfully undergoing an audit, and as of May 1, 2013, 18 of approximately 23 tantalum smelting companies had been certified as conflict-free. As of May 1, 2013, 5 tin smelting companies had been certified as conflict-free, 7 tungsten smelting companies had begun discussions with representatives of the program, and 12 gold refining companies had been certified as conflict-free through the program.

World Gold Council's Conflict-Free Gold Standard and Tools

The World Gold Council developed and issued the Conflict-Free Gold Standard, an industry-led approach to combat the potential misuse of mined gold to fund armed conflict, in October 2012. The standard was developed with council member companies, which constituted the world's leading gold producers, and with extensive input from stakeholders to establish a common approach by which gold producers can assess and provide assurance that their gold has been extracted in a manner that does not cause, support, or benefit unlawful armed conflict or contribute to serious human rights abuses or breaches of international humanitarian law. According to a World Gold Council official, the participating companies' conformance to the Standard will be externally audited and assured and will operationalize the requirements of OECD guidance. The results of the audit using the standard will be recognized across other stakeholder initiatives such as the London Bullion Market Association's Responsible Gold Guidance. The Standard should also support refiners in meeting their due diligence requirements.

In-Region Sourcing Initiatives

Conflict-Free Tin Initiative

The Conflict-Free Tin Initiative (CFTI) is a pilot that was launched in September 2012 and aims to create demand for conflict-free tin from eastern DRC. The traceability and due diligence mechanism through the ITRI Tin Supply Chain Initiative is operated by Pact, an independent NGO, and is operated out of the Kalimbi mine in South Kivu. According to an NGO, the Netherlands Ministry of Foreign Affairs is a neutral broker that brought the partners along the supply chain together, from mine to smelter to end user. The

DRC government and local civil society are closely involved in the initiative, which is structured within the framework of the International Conference of the Great Lakes Region (ICGLR) and will be consistent with the due diligence guidance of OECD. CFTI reported that between October 2012 and January 2013, 210 tons of materials were produced in the Kalimbi mine and the first container of conflict-free tin was transported to the trader in the DRC in December 2012. In January 2013, the first two containers of conflict-free tin were shipped to the smelter in Malaysia. The CFTI reports that next steps will involve the conflict-free tin making its way from the smelter to soldering companies and eventually to end users as finished product.

ICGLR's Regional Certification Mechanism and Other Initiatives

The ICGLR started working with an NGO in 2010 to develop a regional certification mechanism to ensure that conflict minerals are fully traceable. ICGLR's regional certification mechanism may enable member countries and their mining companies to demonstrate where and under what conditions minerals were produced; through the regional certification mechanism, individual member governments are to issue ICGLR regional certificates for those mineral shipments that are in compliance with the standards of the mechanism. According to an official at a partnering NGO, the first two certificates out of the region were scheduled to come from sites in Rwanda and DRC in the late spring and from Uganda by December 2013. However, State indicated that the certificates from Rwanda and DRC have been delayed and will likely not be issued until late summer 2013. Regional certificates from other ICGLR countries will take some time because of capacity issues.

According to USAID, in addition to the regional certification mechanism, ICGLR's other initiatives focused on eliminating the illegal exploitation of natural resources include harmonization of national legislation, formalization of the artisanal mining sector, formalization of the extractives industries transparency initiative, a whistleblowing mechanism, and a regional database on the flow of minerals.

End Notes

[1] GAO, The Democratic Republic of the Congo: U.S. Agencies Should Take Further Actions to Contribute to the Effective Regulation and Control of the Minerals Trade in Eastern Democratic Republic of the Congo, GAO-10-1030 (Washington, D.C.: Sept. 30, 2010).

[2] According to the UN, M23, for March 23 Movement, is an armed group that formed in May 2012 in eastern DRC.

[3] The Dodd-Frank Act defines conflict minerals as columbite-tantalite (coltan), cassiterite, gold, wolframite, or their derivatives, and any other mineral or its derivatives determined by the Secretary of State. Columbite-tantalite (coltan), cassiterite, and wolframite ores are ores from which tantalum, tin, and tungsten, respectively, are processed.

[4] The countries adjoining the DRC are Angola, Zambia, Tanzania, Burundi, Rwanda, Uganda, South Sudan, Central African Republic, and the Republic of the Congo.

[5] 77 Federal Register at 56274 (Sept. 12, 2012).

[6] GAO, The Democratic Republic of the Congo: Information on the Rate of Sexual Violence in War-Torn Eastern DRC and Adjoining Countries, GAO-11-702 (Washington, D.C.: July 13, 2011).

[7] GAO, Conflict Minerals Disclosure Rule: SEC's Actions and Stakeholder-Developed Initiatives, GAO-12-763 (Washington, D.C.: July 16, 2012). In the report, we addressed the mandate requiring us to describe any issues that SEC encountered in promulgating a conflict minerals disclosure rule.

[8] We did not have access to information about specific entities or companies so we described what we learned about entities that are not required to report to the SEC under the conflict minerals rule.

[9] As we reported in 2012, the neighboring countries of eastern DRC are Burundi, Rwanda, and Uganda.

[10] For our review, we selected smelters to interview based on the types of minerals the smelters processed and their willingness to speak with us. We sought to speak with officials of at least one smelter or refiner for each of the conflict minerals—tin, tantalum, tungsten, and gold.

[11] MONUSCO took over from an earlier UN peacekeeping operation—the United Nations Organization Mission in Democratic Republic of the Congo—on July 1, 2010.

[12] Pub. L. No. 109-456, sec 102(14).

[13] Pub. L. No.111-203, sec 1502. According to an expert in conflict minerals that we interviewed, Canada and the European Union have taken steps to propose conflict minerals initiatives: Canada has a proposed conflict minerals rule that is currently in its comment period, and the European Union currently is having consultations on a potential conflicts minerals rule.

[14] According to SEC, when SEC proposes or adopts a set of rules, often those rules are contained in a single document, called a proposing release or adopting release.

[15] In 2012 we reported on some of the factors that caused delays in developing, modifying, and finalizing the rule. GAO-12-763.

[16] SEC reported that commentary came from corporations, professional associations, human rights and public policy groups, bar associations, auditors, institutional investors, investment firms, U.S. and foreign government officials, consumers, and other interested parties and stakeholders.

[17] 15 U.S.C. § 78m(a) and 78o(d).

[18] According to SEC, the report will be filed on Form SD.

[19] According to the SEC rule, SEC-reporting companies may describe their products as "DRC conflict undeterminable" if, following their exercise of due diligence, they are unable to determine that their minerals meet the statutory definition of "DRC conflict-free" because they have conflict minerals from the DRC or the adjoining countries but were unable to determine if their conflict minerals financed or benefited armed groups; or they had reason to believe that their conflict minerals may have originated in the DRC and adjoining countries and may not have come from recycled or scrap sources and the information

gathered failed to clarify the conflict minerals' country of origin, benefits to armed groups, or whether it came from scrap or recycled sources.

[20] National Association of Manufacturers, Chamber of Commerce of the United States of America, and Business Roundtable v. U.S. Securities and Exchange Commission, 12- 1422, U.S. Court of Appeals for the District of Columbia Circuit. In May 2013, the U.S. Court of Appeals transferred the case to the U.S. District Court for the District of Columbia citing a recent jurisdictional decision based on a similar case.

[21] See GAO-12-763. For example, tin solder is used to attach individual components on circuit boards.

[22] Tantalum is also used to produce alloy additives, which can be found in turbines in jet engines; mill and chemical products; thin films, which are used in semiconductors; and other products.

[23] Artisanal mining is a form of mining that is characterized by a lack of mechanization or capital investment.

[24] According to OECD, in many instances smelters and refiners do not actually take ownership of the mineral but provide a service and charge a fee based on the amount of minerals smelted. Ownership of the minerals may remain with the mineral trader, bank, or component parts manufacturer.

[25] According to a World Gold Council representative, in most cases refiners are paid a fee to refine gold and the transaction is conducted between the miner, trader, and the bank.

[26] OECD defines "due diligence" as an ongoing, proactive, and reactive process through which companies can ensure that they respect human rights and do not contribute to conflict. Due diligence can also help companies ensure that they observe international law and comply with domestic laws, including those governing the illicit trade in minerals and UN sanctions. Risk-based due diligence refers to the steps companies should take to identify and address actual or potential risks in order to prevent or mitigate adverse impacts associated with their activities or sourcing decisions.

[27] Some agency officials and NGOs mentioned that the uncertainty of the SEC rule's outcome due to it being challenged may make it harder for companies to move forward in implementing their due diligence process to address the SEC reporting requirements.

[28] The World Bank defines the term "civil society" to refer to the wide array of nongovernmental and not-for-profit organizations that have a presence in public life, expressing the interests and values of their members or others, based on ethical, cultural, political, scientific, religious, or philanthropic considerations.

[29] Smelters are the natural choke point in the supply chain—meaning that there are numerous sources of raw materials (ore) that flow into a smelter and numerous uses for the refined metals that leave the smelter.

[30] According to industry representatives, independent third-party audits for the Conflict-Free Smelter Program are lagging and thus cannot assure that minerals are conflict-free on future purchasing actions.

[31] GAO-10-1030.

[32] The list of factors is not intended to be exhaustive and is based on views of experts and stakeholders. Additional factors may exist once the SEC rule has been in place for a longer period of time. We did not address how the challenges raised by the factors discussed can be mitigated or what progress the DRC government has made to address them because it was outside the scope of our review.

[33] Organisation for Economic Co-operation and Development, Upstream Implementation of the OECD Due Diligence Guidance for Responsible Supply Chains of Minerals from Conflict-

Affected and High-Risk Areas: Final Report on One-Year Pilot Implementation of the Supplement on Tin, Tantalum, and Tungsten (OECD: January 2013).

[34] Letter dated 12 November 2012 from the Chair of the Security Council to resolution 1533 (2004) concerning the Democratic Republic of the Congo addressed to the President of the Security Council (S/2012/843), the United Nations Security Council (Nov. 15, 2012).

[35] The actual number of first-tier suppliers is unknown, and organizations that commented on the release provided varying estimates on the number of first-tier suppliers that could potentially be affected by the SEC rule. As cited in the SEC release, commentators provided estimates indicating that SEC-reporting companies under the rule averaged nearly 160 to 10,000 first-tier suppliers, each. In addition, another commentator estimated that SEC-reporting companies averaged 1,060 first-tier suppliers each. After accounting for redundancies, because a supplier may be in more than one supply chain, SEC revised the number of potentially affected suppliers, provided by one commentator, to 278,000 suppliers.

[36] For the purpose of the estimate, SEC estimated that the rule would affect 5,551 of 13,545 domestic issuers that file form 10-K, 377 of 942 foreign private issuers that file form 20-F, and 66 of 205 Canadian issuers that file form 40-F, annually. According to the adopting release for the rule, SEC based its estimate of 5,994 affected companies on the number of issuers that fall under all Standard Industrial Classification (SIC) codes that SEC staff believed were most likely to manufacture or contract to manufacture products with necessary conflict minerals.

[37] SEC estimates that a number of SEC-reporting companies under the rule may be original equipment manufacturers and first-tier suppliers, although SEC-reporting companies under the rule could exist anywhere on the supply chain.

[38] According to EICC and GeSI, as of April 30, 2013, the EICC and GeSI activities are now under the Conflict Free Sourcing Initiative, which is an expanded initiative of the EICC and GeSI Extractives Work Group that includes more stakeholders and a wider range of industry sectors supporting the sourcing of conflict-free minerals.

[39] According to SEC officials, since most of these companies do not file reports with SEC under Section 13(a) or 15(d) of the Securities Exchange Act, they would not be required to report under the rule.

[40] We are reporting information we found and analyzed on smelters and refiners to address the mandated question concerning available information about entities that use conflict minerals and do not report to SEC under the rule.

[41] Five additional tin smelters have successfully completed a Conflict Free Smelter Program audit, bringing the number of certified “conflict-free” smelters to 35; however, the EICC and GeSI has yet to identify these smelters publicly, including the location of the smelters, due to binding non-disclosure agreements with the smelters.

[42] LMBA created the Responsible Gold Guidance for Good Delivery Refiners to combat systematic or widespread abuses of human rights, and to avoid contributing to conflict, among other abuses, according to the guidance. In addition, the guidance follows the five steps framework for risk-based due diligence of the OECD Due Diligence Guidance for Responsible Supply Chains of Minerals from Conflict-Affected and High-Risk Areas adopted on December 15, 2010 and follows the requirements detailed in the OECD Gold Supplement adopted on July 17, 2012. Since January 2012, the LBMA has required all Good Delivery Gold Refiners to comply with the Responsible Gold Guidance.

[43] We were unable to identify a website for 86 of the 278 smelters in our analysis, and therefore were unable to determine whether these companies have policies regarding procuring

conflict minerals from the DRC and adjoining countries. The inability to locate some websites of smelters in our analysis could be linked to factors such as language as well as company names and websites that may contain characters not found in the Roman alphabet. According to an official with the EICC and GeSI, some smelters may have such small operations at the mine site in remote locations that they do not create websites.

[44] S olutions for Hope is a "closed-pipeline" initiative to trace the flow of tantalum from the mine to the end-use company.

[45] GAO-12-763.

[46] United Nations Security Council, Sexual Violence in Conflict: Report of the Secretary-General, A/67/792-S/2013/149 (New York, NY: Mar 14, 2013).

End Note for Appendix I

[1] GAO, Conflict Minerals Disclosure Rule: SEC's Actions and Stakeholder-Developed Initiatives, GAO-12-763 (Washington, D.C.: July 16, 2012). We did not include information on studies or reports that were preliminary.

End Note for Appendix III

[1] We did not conduct a comprehensive review of all the initiatives we identified in our 2012 report since that was not within the scope of this review. In the process of this review, we did obtain some new information pertaining to selected global and in-region sourcing initiatives, which is reflected in this appendix. However, we did not talk to or obtain information from the government of the DRC on in-region initiatives or legislation related to conflict minerals because this was outside of the scope of our current review.

[2] Organisation for Economic Co-operation and Development, OECD Due Diligence Guidance for Responsible Supply Chains of Minerals from Conflict-Affected and High-Risk Areas, second edition (OECD Publishing, 2011). Accessed July 2, 2013, http://dx.doi.org/10.1787/9789264185050-en. The guidance was adopted on May 25, 2011.

[3] Organisation for Economic Co-operation and Development, Downstream Implementation of the OECD Due Diligence Guidance for Responsible Supply Chains of Minerals from Conflict-Affected and High-Risk Areas: Final Downstream Report on One-Year Pilot Implementation of the Supplement on Tin, Tantalum, and Tungsten (OECD: January 2013); and Upstream Implementation of the OECD Due Diligence Guidance for Responsible Supply Chains of Minerals from Conflict-Affected and High-Risk Areas: Final Report on One-Year Pilot Implementation of the Supplement on Tin, Tantalum, and Tungsten (OECD: January 2013).

INDEX

S

T